INTEGRATED PEST MANAGEMENT FOR

RICE

THIRD EDITION

PREPARED BY THE
UNIVERSITY OF CALIFORNIA
STATEWIDE INTEGRATED PEST MANAGEMENT PROGRAM

University of California Agriculture and Natural Resources
OAKLAND, CALIFORNIA
PUBLICATION 3280

For information about ordering this publication and/or a free catalog, contact
University of California
Agriculture and Natural Resources
Communication Services
1301 S. 46th Street
Building 478 - MC 3580
Richmond, CA 94804-4600
Telephone 1-800-994-8849
510-665-2195
FAX 510-665-3427
E-mail: anrcatalog@ucanr.edu

Third edition, 2013

Publication 3280
ISBN-13: 978-1-60107-753-0
Photographs by Jack Kelly Clark (except as noted). Cover photo by Jim Morris, California Rice Commission; back cover photo by Mike Poe.

Library of Congress Cataloging-in-Publication Data

Integrated pest management for rice / prepared by the University of California Statewide Integrated Pest Management Program. -- 3rd ed.

p. cm. -- (Publication ; 3280)

Includes bibliographical references and index.

ISBN 978-1-60107-753-0

1. Rice--Diseases and pests--Integrated control--California. 2. Agricultural pests--Integrated control--California. I. University of California Integrated Pest Management Program. II. University of California (System). Division of Agriculture and Natural Resources. III. Series: Publication (University of California (System). Division of Agriculture and Natural Resources) ; 3280.

SB608.R5157 2012

632'.9--dc23

2012028867

This publication has been anonymously peer reviewed for technical accuracy by University of California scientists and other qualified professionals. This review process was managed by the ANR Associate Editor for Pest Management.

Printed in the United States on recycled paper.

750-pr-1/13-LR/RW

WARNING ON THE USE OF CHEMICALS

Pesticides are poisonous. Always read and carefully follow all precautions and safety recommendations given on the container label. Store all chemicals in their original labeled containers in a locked cabinet or shed, away from foods or feeds, and out of the reach of children, unauthorized persons, pets, and livestock.

Recommendations are based on the best information currently available, and treatments based on them should not leave residues exceeding the tolerance established for any particular chemical. Confine chemicals to the area being treated. THE GROWER IS LEGALLY RESPONSIBLE for residues on the grower's crops as well as for problems caused by drift from the grower's property to other properties or crops.

Consult your county agricultural commissioner for correct methods of disposing of leftover spray materials and empty containers. Never burn pesticide containers.

PHYTOTOXICITY: Certain chemicals may cause plant injury if used at the wrong stage of plant development or when temperatures are too high. Injury may also result from excessive amounts or the wrong formulation or from mixing incompatible materials. Inert ingredients, such as wetters, spreaders, emulsifiers, diluents, and solvents, can cause plant injury. Since formulations are often changed by manufacturers, it is possible that plant injury may occur, even though no injury was noted in previous seasons.

PRECAUTIONS FOR USING PESTICIDES (IPM)

Pesticides are poisonous and must be used with caution. READ THE LABEL CAREFULLY BEFORE OPENING A PESTICIDE CONTAINER. Follow all label precautions and directions, including requirements for protective equipment. Use a pesticide only on crops specified on the label. Apply pesticides at the rates specified on the label or at lower rates if suggested in this publication. In California, all agricultural uses of pesticides must be reported. Contact your county agricultural commissioner for details. Laws, regulations, and information concerning pesticides change frequently, so be sure the publication you are using is up to date.

Legal Responsibility. The user is legally responsible for any damage due to misuse of pesticides. Responsibility extends to effects caused by drift, runoff, or residues.

Transportation. Do not ship or carry pesticides together with foods or feeds in a way that allows contamination of the edible items. Never transport pesticides in a closed passenger vehicle or in a closed cab.

Storage. Keep pesticides in original containers until used. Store them in a locked cabinet, building, or fenced area where they are not accessible to children, unauthorized persons, pets, or livestock. DO NOT store pesticides with foods, feeds, fertilizers, or other materials that may become contaminated by the pesticides.

Container Disposal. Dispose of empty containers carefully. Never reuse them. Make sure empty containers are not accessible to children or animals. Never dispose of containers where they may contaminate water supplies or natural waterways. Consult your county agricultural commissioner for correct procedures for handling and disposal of large quantities of empty containers.

Protection of Nonpest Animals and Plants. Many pesticides are toxic to useful or desirable animals, including honey bees, natural enemies, fish, domestic animals, and birds. Crops and other plants may also be damaged by misapplied pesticides. Take precautions to protect nonpest species from direct exposure to pesticides and from contamination due to drift, runoff, or residues. Certain rodenticides may pose a special hazard to animals that eat poisoned rodents.

Posting Treated Fields. For some materials, reentry intervals are established to protect field workers. Keep workers out of the field for the required time after application and, when required by regulations, post the treated areas with signs indicating the safe reentry date.

Harvest Intervals. Some materials or rates cannot be used in certain crops within a specific time before harvest. Follow pesticide label instructions and allow the required time between application and harvest.

Permit Requirements. Many pesticides require a permit from the county agricultural commissioner before possession or use. When such materials are recommended in this publication, they are marked with an asterisk (*).

Processed Crops. Some processors will not accept a crop treated with certain chemicals. If your crop is going to a processor, be sure to check with the processor before applying a pesticide.

Crop Injury. Certain chemicals may cause injury to crops (phytotoxicity) under certain conditions. Always consult the label for limitations. Before applying any pesticide, take into account the stage of plant development, the soil type and condition, the temperature, moisture, and wind direction. Injury may also result from the use of incompatible materials.

Personal Safety. Follow label directions carefully. Avoid splashing, spilling, leaks, spray drift, and contamination of clothing. NEVER eat, smoke, drink, or chew while using pesticides. Provide for emergency medical care IN ADVANCE as required by regulation.

Other books in this series include:
Integrated Pest Management for Alfalfa Hay, Publication 3312
Integrated Pest Management for Almonds, Second Edition, Publication 3308
Integrated Pest Management for Apples and Pears, Second Edition, Publication 3340
Integrated Pest Management for Avocados, Publication 3503
Integrated Pest Management for Citrus, Third Edition, Publication 3303
Integrated Pest Management for Cole Crops and Lettuce, Publication 3307
Integrated Pest Management for Cotton, Second Edition, Publication 3305
Integrated Pest Management for Floriculture and Nurseries, Publication 3402
Integrated Pest Management for Potatoes, Second Edition, Publication 3316
Integrated Pest Management for Small Grains, Publication 3333
Integrated Pest Management for Stone Fruits, Publication 3389
Integrated Pest Management for Strawberries, Second Edition, Publication 3351
Integrated Pest Management for Tomatoes, Fourth Edition, Publication 3274
Integrated Pest Management for Walnuts, Third Edition, Publication 3270
Natural Enemies Handbook: The Illustrated Guide to Biological Pest Control, Publication 3386
Pests of Landscape Trees and Shrubs, Second Edition, Publication 3359
Pests of the Garden and Small Farm, Second Edition, Publication 3332

Contributors and Acknowledgments

Third edition written by Larry L. Strand, Principal Editor

Supervising Editor: Tunyalee A. Martin, Content Development Supervisor

Cover photo by Jim Morris, California Rice Commission; back cover photo by Mike Poe. All other photographs by Jack Kelly Clark (except as noted).

Prepared by the University of California Statewide IPM Program at Davis.

Technical Editors for Third Edition

Luis A. Espino, University of California Cooperative Extension, Colusa

Albert J. Fischer, Department of Plant Sciences, University of California, Davis

Larry D. Godfrey, Department of Entomology, University of California, Davis

Christopher A. Greer, University of California Cooperative Extension, Yuba City

James E. Hill, Department of Plant Sciences, University of California, Davis

Rex E. Marsh, Department of Wildlife, Fish, and Conservation Biology, University of California, Davis

Randall G. Mutters, University of California Cooperative Extension, Oroville

Contributors to Third Edition

The following have generously provided information, offered suggestions, reviewed draft manuscripts, or helped obtain photographs (University of California, except as noted).

Entomology: Luis A. Espino, Larry D. Godfrey, Sharon P. Lawler, Michael J. Stout (Louisiana State University)

Horticulture, Physiology, Soil and Water Relations: Merle Anders (University of Arkansas), James E. Hill, Randall G. Mutters, Lloyd T. Wilson (Texas A&M University)

Plant Pathology: Christopher A. Greer, Donald E. Groth (Louisiana State University)

Vertebrate Biology: W. Paul Gorenzel, John Hare (Wilbur-Ellis), Rex E. Marsh

Weed Science: Ellen A. Dean, James W. Eckert, Albert J. Fischer, David F. Spencer, Eric P. Webster (Louisiana State University)

Special Thanks

We also acknowledge the important role of the contributors to the first and second editions of this manual, which were published in 1983 and 1993: David E. Bayer, D. Marlin Brandon, Richard W. DeHaven, Albert A. Grigarick, Milton D. Miller, Duane S. Mikkelsen, Maurice L. Peterson, Terrell P. Salmon, Steven C. Scardaci, Robert K. Washino, Robert K. Webster, Carl M. Wick, and John F. Williams.

Information in this manual has been derived from research supported by the University of California, the California Rice Research Board, and the rice industry.

Production

Production (third edition): Robin Walton, ANR Communication Services

Design (IPM manual series): Seventeenth Street Studios

Drawings (third edition): Robin Walton, based on second edition

Drawings (first and second editions): Marvin Ehrlich and David Kidd

Editing (third edition): Linda Ribera, ANR Communication Services

Contents

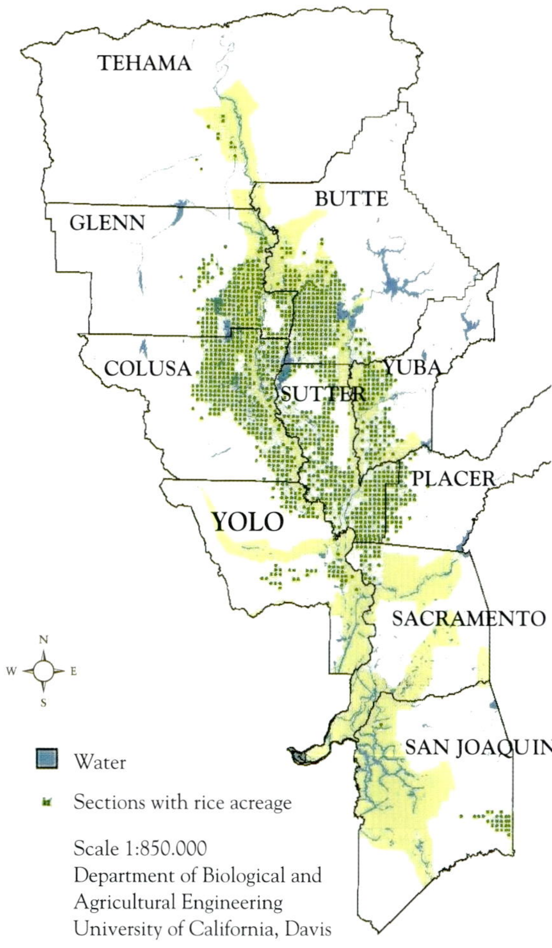

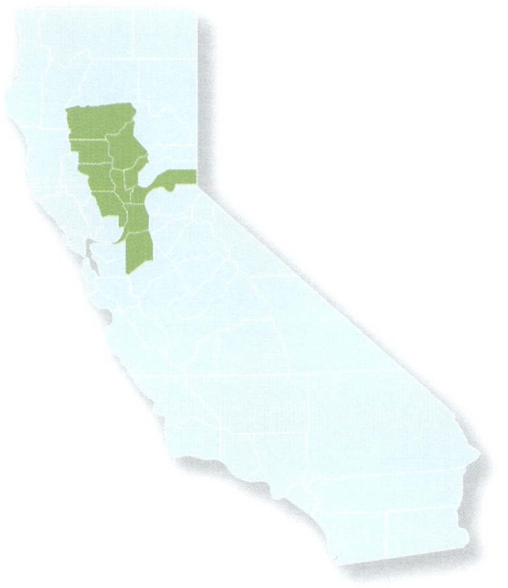

Figure 1. Regions of California where rice is grown.

Integrated Pest Management for Rice

Integrated pest management (IPM) provides a long-term strategy for minimizing losses caused by pests with as little cost to the grower and disruption of the environment as possible. This manual presents IPM methods and strategies for use in rice.

The foundation of the California rice industry includes the availability of clean, abundant, low-cost water and land well suited to rice production. Rice is grown annually on about 450,000 to 550,000 acres in California; about 95% of this acreage is concentrated in the Sacramento Valley, with the balance scattered throughout the San Joaquin Valley (Fig. 1). California rice varieties and many of the agronomic practices are quite different from those of other rice-producing states because of the dry Mediterranean climate and northern latitude. Over 90% of the area is planted to medium-grain rice, with limited planting of short- and long-grain varieties. Seasonal length for the majority of early-maturing varieties is 140 to 145 days. The irrigation season is about 120 days. Postharvest activities include management of straw residue, primarily by soil incorporation; there is a limited amount of burning for disease control and a small amount of removal for use in straw-based products.

Much of the land devoted to rice production in California is poorly suited for production of other crops, primarily because of impeded internal drainage. The fine-textured soils in the basin areas of the Sacramento Valley were wetlands formed by the annual overflow of valley streams and rivers; older soils on the valley's edges have dense hardpan or claypan subsoils. Both types of soil are well suited to rice production because their low percolation rates efficiently maintain the floodwater. Conversion to rice culture has improved some of these difficult soils with leveling and surface drainage so that alternative crops can now be grown in some areas. However, on a substantial portion of land where rice is grown, rice is the only profitable crop that can be produced.

The climate in California rice-producing areas includes a long growing season that provides ample time for the crop to mature. Plentiful sunlight and adequate heat ensure timely growth and development, and dry weather minimizes foliar diseases. However, the relatively cool night temperature during flowering has made it necessary to select for more cold-tolerant cultivars to reduce the incidence of cold-induced blanking (lack of seed set). Best crop performance is associated with moderate temperatures throughout the growing season.

A plentiful supply of high-quality water is necessary for the economic production of rice. With two major river systems, the Sacramento and Feather, and their associated reservoirs and tributaries, the Sacramento Valley usually has a sufficient supply of water to sustain efficient production of rice. However, environmental changes may limit the availability of water in the future. Less rice is grown in the San Joaquin Valley primarily because the water is more expensive and less available, and higher-value crops can be grown.

Approximately 90% of California rice is irrigated with surface water. Consumptive use (evapotranspiration) of water by rice is equivalent to that of several other crops, including alfalfa, irrigated pasture, and deciduous orchards. The approximate seasonal water use as evapotranspiration (evaporation plus transpiration) is about 3.5 acre-feet (4,317 cu m). There is extensive reuse of drainage water within irrigation districts and on-farm recirculation systems. Deep percolation is minimized by avoiding fields with high infiltration rates.

In their original swampy condition, most of the rice-growing areas provided a good habitat for many aquatic plants and animals. Some of these native inhabitants have adapted to rice culture and become serious pests, providing rice with a pest fauna and flora that is unique among California's agricultural crops. On the other hand, rice culture provides a favorable environment for a number of desirable wildlife species. An introduced species, the pheasant, and several species of migratory waterfowl have adapted to the rice environment. The practice of winter flooding has increased the populations of waterfowl that overwinter in rice-growing areas.

Over time, specialized strategies have been developed to manage rice culture's unusual problems. The industry has been highly successful in developing partnerships with urban and conservation groups, but these compromises have transformed relatively simple agronomic practices into a complicated, management-intensive crop production system.

All California rice is direct-seeded; most is pregerminated and aerially seeded into standing water under a continuous flood. The water-seeded system was established to suppress watergrasses, *Echinochloa* spp. Increasingly, rice is drill-seeded or dry-seeded and permanently flooded after stand establishment. Flooded conditions create a more favorable environment for rice growth and development by moderating temperature extremes and improving the availability of soil phosphorus. The flooded conditions also protect developing seeds and seedlings from most vertebrates. However, flooded fields provide an environment favorable for mosquitoes. They are controlled by the release and culture of a mosquito-eating fish, *Gambusia affinis*, and the use of the microbial insecticide *Bacillus thuringiensis israelensis*.

In some situations, rice is planted into dry seedbeds (dry-seeding). Dry-seeded fields are intermittently irrigated prior to being flooded after stand establishment. This technique is used to reduce production costs or to control herbicide-resistant weed species that do not thrive in dry-seeded fields. Some pests, such as aquatic weeds and tadpole shrimp, are much less of a problem in dry-seeded fields, but the watergrasses, *Echinochloa* spp., are more of a problem.

Pest control practices in rice have generated some environmental concerns. One concern is the downstream movement of irrigation water that may contain pesticide residues. A combination of regulations, improved irrigation system design, and more efficient water management has greatly mitigated these problems and reduced herbicide runoff from rice fields by more than 95%. The rice industry continues to work with the Regional Water Quality Control Board and other agencies to further reduce the impact of rice pesticides on water quality. A second concern involves drift of aerially applied herbicides, which may damage neighboring orchards and crops. Over the years, drift problems have been minimized by herbicide-use restrictions, increased use of ground application, the development of more-precise nozzles and application equipment, and improved formulations and additives. The practice of burning rice fields, a technique that reduces toxic gas and stem rot disease problems, is also a concern. An alternative to burning, removal or incorporation of rice straw followed by winter flooding, is now used for most rice fields.

Information about the latest production practices can be found in the most recent *California Rice Production Workshop Workbook*, listed in the suggested reading, and on the University of California Rice Research and Information Program website (www.plantsciences.ucdavis.edu/uccerice/). *Integrated Pest Management for Rice* is designed to help you apply IPM principles in the management of a rice crop; it is not intended to be a complete production guide.

Read the manual's sections on crop growth and development and general management practices for background information on the reasoning behind IPM strategies. In the chapter "Managing Pests in Rice," the chart of management considerations will help you plan your IPM program and predict or prevent potential problems before they occur. Use the photographs and descriptions in the pest sections to help identify pests and pest damage. Check the glossary for definitions of unfamiliar terms.

Read the biological information and management overview, but consult *UC IPM Pest Management Guidelines: Rice*, listed in the suggested reading and available at www.ipm.ucdavis.edu, before taking control actions. The guidelines contain specific, current information on registered pesticides and other pest control tools. Because pesticide registrations change frequently, it is not possible to provide complete information on application rates and current registrations in this manual.

Research on individual pest problems in rice has been extensive and productive, and California researchers are continuing to develop comprehensive interdisciplinary IPM programs. The information presented in this manual is a composite of research, biological information, and the experience of experts in the field. Scientists at the University of California, USDA, and the Rice Experiment Station are continuing to develop new rice varieties, cultural methods, fertilizer practices, monitoring techniques, pesticide recommendations, and other methods that can be combined in an IPM program. Also check with your farm advisor or area IPM advisor to find out about new developments.

The Rice Plant: Development and Growth Requirements

Knowledge of basic plant biology is essential for planning a successful integrated pest management program. It is especially important to understand how the environment interacts with pests and cultural practices to affect rice development and yields. Without this understanding, it is easy to overlook stress symptoms or to confuse them with pest damage. Cultural practices and pest control methods should be evaluated for their impact on the total crop system as well as for their effectiveness in eliminating the primary problem. Otherwise, these practices may compound the stress on the crop and add to economic losses.

Rice yield (Y) is a product of four yield components: the number of panicles (flowering heads) per unit area (P), the number of spikelets per panicle (S), the proportion of spikelets that become filled grains (G), and the weight of the single grains (W):

$$Y = P \times S \times G \times W$$

Pest problems, environmental conditions, and management practices can affect all of these yield components. Rather than focusing solely on reducing damage by certain pests, a good IPM program aims to maximize these yield components.

Stages of Crop Development

Rice proceeds through three main phases of development: the vegetative phase, from seed germination to panicle initiation; the reproductive phase, from panicle initiation to flowering; and the ripening phase, from flowering to grain maturity. The time it takes to pass through these stages is largely controlled by temperature and the rice cultivar (Fig. 2); for example, early-maturing cultivars spend fewer days in the vegetative phase. The rate of development is also dependent on air and water temperatures, day length, seeding rate, and fertilization rate, and how the field is seeded during planting. For example, dry-seeded rice takes longer, about 7 additional days or more, to reach full grain maturity than does water-seeded or presoaked, water-seeded rice. Management considerations and pest problems differ for each growth stage (see Table 1 in the next chapter). Growth stages are illustrated in Figure 3.

The Vegetative Phase

Soaking and Draining Seed. Seed to be planted in flooded fields are soaked in water 18 to 24 hours to pregerminate them

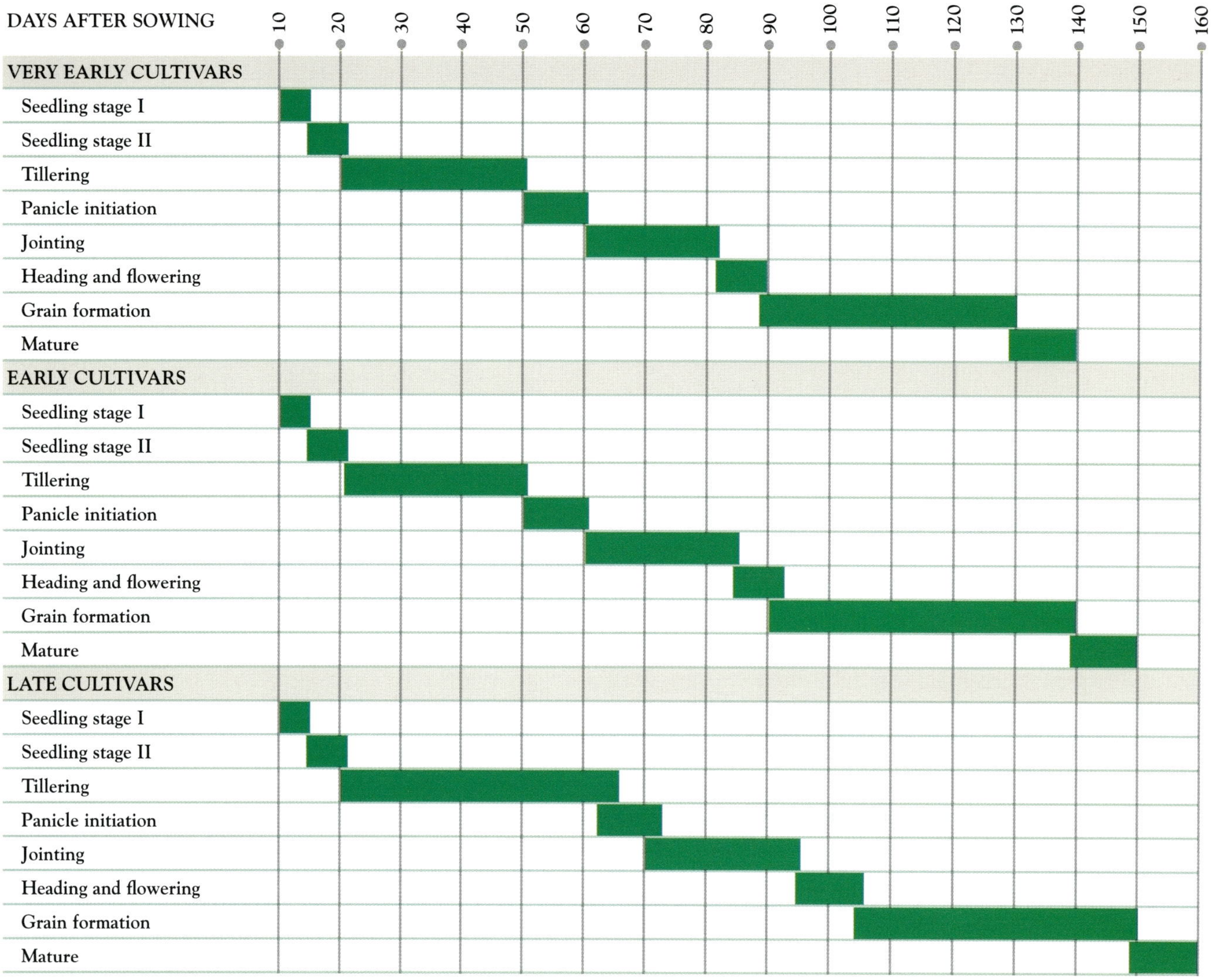

DAYS AFTER SOWING
10
20
30
40
50
60
70
80
90
100
110
120
130
140
150
160
VERY EARLY CULTIVARS
Seedling stage I
Seedling stage II
Tillering
Panicle initiation
Jointing
Heading and flowering
Grain formation
Mature
EARLY CULTIVARS
Seedling stage I
Seedling stage II
Tillering
Panicle initiation
Jointing
Heading and flowering
Grain formation
Mature
LATE CULTIVARS
Seedling stage I
Seedling stage II
Tillering
Panicle initiation
Jointing
Heading and flowering
Grain formation
Mature

Figure 2. Stages of rice development and their approximate duration in early, very early, and late rice cultivars grown under normal California conditions.

Tiller initiation
Germination
Preflood
Early tillering
Midtillering
Panicle initiation
Flowering
Days: early variety
25–30
55–60
85–90
Days: late variety
25–30
65–75
95–105

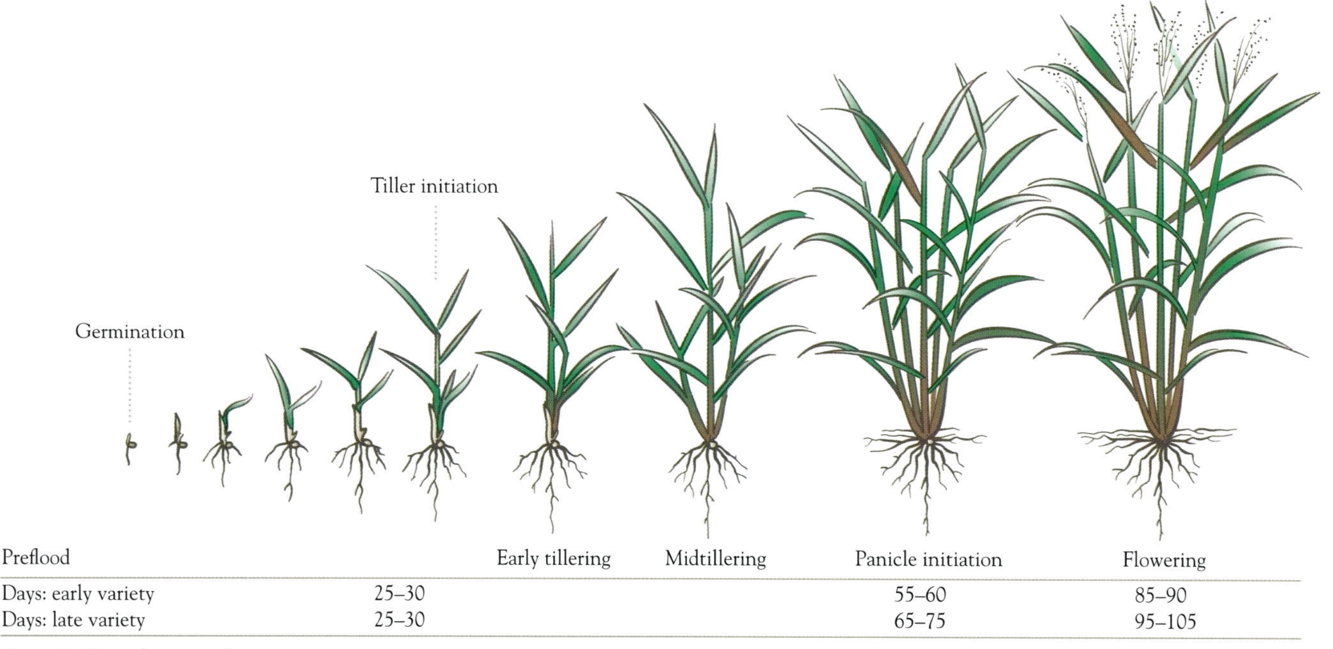

Figure 3. Growth stages of rice.

(i.e., germinate them before planting to help achieve faster and better stand establishment) and to remove the air inside the hull, which reduces seed floating. After the water is drained from the soaking containers, seed should be planted about 24 to 48 hours later. Draining accelerates embryo development by exposing the seed to oxygen and allowing the seed rather than water to absorb most of the heat released in the germination process. However, heat from respiration can rapidly increase to damaging levels and can increase the levels of the bakanae pathogen; if planting is delayed, cool the seed with fresh water. Draining also reduces clumping to ensure uniform distribution of the seed when sown from the air. Seed to be dry-planted should not be pregerminated.

Seedling Stage I: From Germination to 3 Leaves. During the first 10 days to 2 weeks after sowing, the seedling is largely dependent upon energy stored as starch in the seed. The first structure to appear is the membranous seed leaf, or coleoptile, then the primary root (radicle), followed by the first foliage leaf, which has a sheath and no blade, and subsequent leaves with blades (Fig. 4). The coleoptile first appears as a tubular structure that splits lengthwise and curls outward, allowing the first foliage leaf to emerge through it. The rice leaf consists of a flat blade and a tubular sheath that surrounds the stem (culm). The joint where the blade attaches to the sheath is called the collar. A ligule and auricles are present at the collar (see Fig. 11 in the chapter "Weeds"). Each new leaf emerges from within the sheath of the preceding leaf. A system of cells called aerenchyma forms concurrently with the third leaf; aerenchyma is a cavity-filled tissue that allows oxygen to be transported from the air through stems to the roots. Without aerenchyma, rice produced in flooded environments would suffer from oxygen deficiency and would be much less vigorous in its growth. Under water-sown conditions, the shoot develops more rapidly than the root system until the shoot emerges through the water. Under dry-seeded conditions, the root grows faster than the shoot.

Seedling Stage II: From 4 to 5 Leaves. During the third week after germination, when the fourth leaf is developing, the seedling no longer relies on stored energy because the young seedling can now be sustained from energy developed by photosynthesis. The shoot should emerge above the water surface at about this time. Crop growth will be delayed in fields that are flooded too deeply or that have cloudy water during this stage

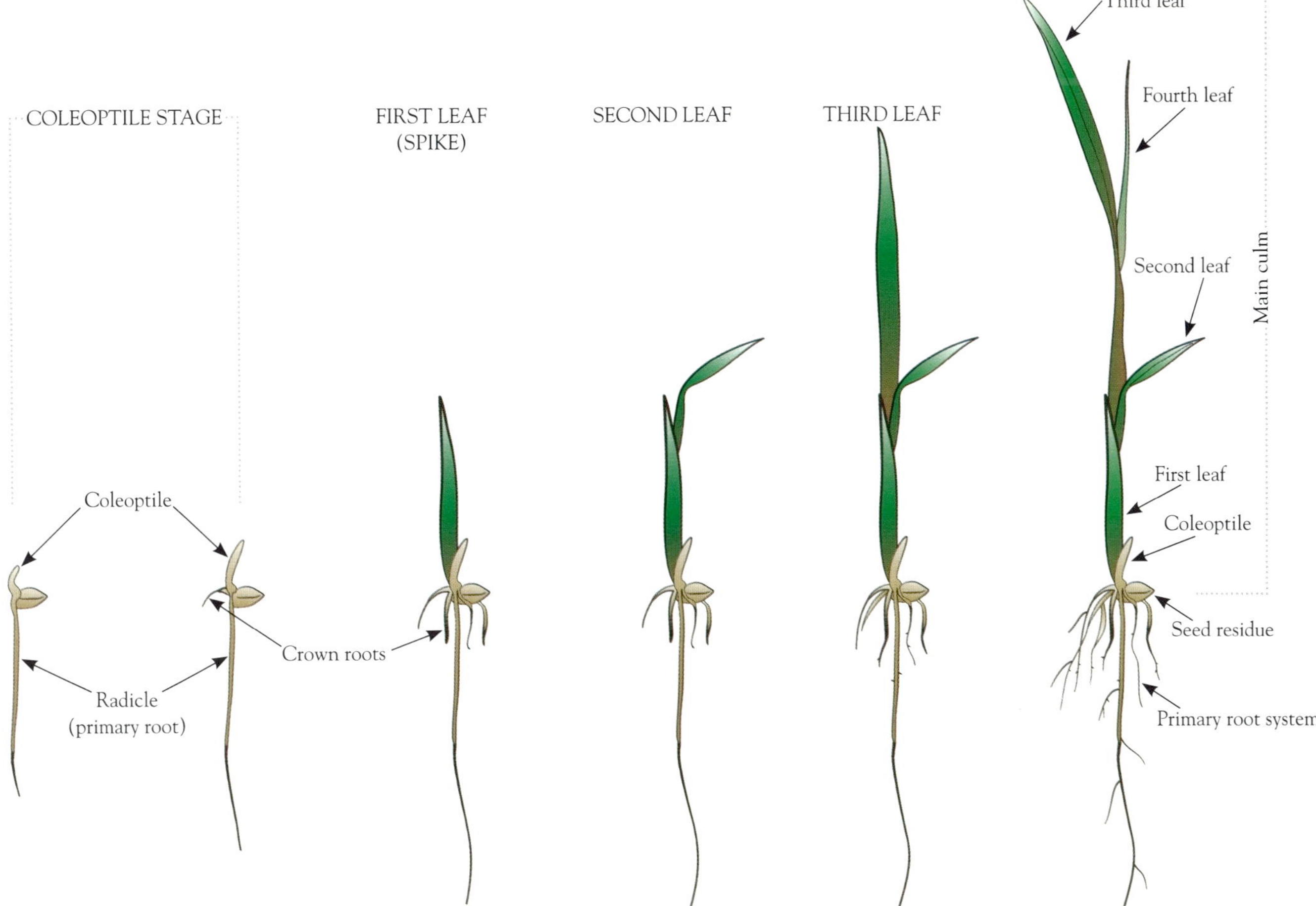

Figure 4. Development of rice seedlings.

Figure 5. First primary tiller arising between base of second leaf and main stem. Each tiller develops adventitious roots, leaves, a stem, and panicle.

of crop development. The root system continues to expand and supplies the plant with nutrients from the soil. The fifth leaf develops, and the first tiller buds form.

Tillering. At tillering, primary tillers develop in the axils at the base of each leaf. Tillers typically begin to appear about the fifth leaf stage. The first primary tiller usually develops at the base of the second leaf (Fig. 5), and additional tillers may arise from the third, fourth, and subsequent leaf axils. Tillers arising from the main culm are primary; those arising from primary tillers are secondary. Under favorable conditions, most tillers developing during the first 7 to 8 weeks after planting will develop panicles before harvest. Tillering is one of the critical stages that can be most influenced by management practices. The total number of tillers produced is regulated primarily by plant population and soil fertility, but cultivar, planting time, and stresses caused by pests and environmental conditions can also affect tiller production. The number of tillers forming at this time determines the potential number of panicles produced, an important yield component. However, with seeding rates of 150 pounds (67.5 kg) or more per acre commonly used in California, stands are so dense that each plant usually produces only two to four productive tillers.

Maximum Tillering. The point in time when the rice plant has formed all the tillers it is going to produce is called maximum tillering, and it occurs before the end of the vegetative growth phase. By this time the crown roots, which form the primary root system, are fully developed. Typically, final tiller number is less than maximum tiller number because some tillers stop developing and senesce (turn yellow and die). This reduction in tiller number is more pronounced in shallow water, which promotes the development of many nonreproductive tillers, than in deep water, where the maximum tiller number and final tiller number are nearly equal. The final tiller number can be lower in deep-water than in shallow-water rice. However, yield potentials are comparable in the two systems if adequate plant numbers are present in both.

Following tillering, the growth rate slows for a short period, and the plant may temporarily yellow as it changes from vegetative to reproductive growth. Temperature and cultivar determine when the plant enters its reproductive phase. For late-season cultivars, day length also influences when the reproductive phase begins.

The Reproductive Phase

Panicle Initiation. The panicle begins to form within the culm and leaf sheath at the base of the tiller, just above the soil surface. After about a week, the embryonic panicle grows to a length of 1⁄25 inch (1mm), just large enough to be visible with the naked eye when you slice open the stem lengthwise where it bulges slightly at the base. At this time, panicle differentiation occurs: cells begin to organize into distinct plant parts. Five days later, the panicle is about ¾ inch (10–20 mm) long. The number of spikelets per panicle is determined during this period. At panicle initiation, adventitious roots begin to develop at or just below the soil surface; these roots form a mat on the soil surface and gradually take over from the crown roots.

Jointing. Five days after panicle initiation, the internodes between the leaves on the top part of the rice plant begin to elongate. The panicle is 1 to 2 inches (2.5–5 cm) above the crown at this time. Internode elongation, known as jointing, starts at the fourth to sixth internode from the top and continues upward. Final internode lengths are longer at the top and smaller toward the bottom of the plant. Elongation continues until the time of flowering.

The panicle, still within the upper leaf sheath and not visible, expands, creating a bulge or boot. During the final stages of

Pollen formation can be inhibited by cold temperatures. The plant in the center, with the collar of the flag leaf and the collar of the previous leaf aligned, is at the most sensitive stage. The other plants illustrate stages just before cold sensitivity (left) and just after (right).

panicle development, pollen is formed within each immature spikelet. Temperatures of 55°F (13°C) or lower at this time can inhibit pollen formation, lowering fertility and causing sterility (blanking). Excess levels of nitrogen fertilizer, especially during cool seasons, can also cause blanking.

The collars of the flag leaf (the last emerging leaf) and the next lower leaf are usually aligned at this cold-sensitive stage. If potentially injurious air temperatures are anticipated, raising the water level above the level of the developing panicle can minimize injury.

Heading and Flowering. At heading, the internodes elongate rapidly, pushing the panicle through the flag leaf sheath (Fig. 6), which is the thirteenth or fourteenth and final leaf for the majority of rice varieties. Flowering occurs about 35 days after panicle initiation and begins near the top of the panicle; it then proceeds downward and from the branch tips inward. The complete panicle flowers in about 7 days; the last florets to

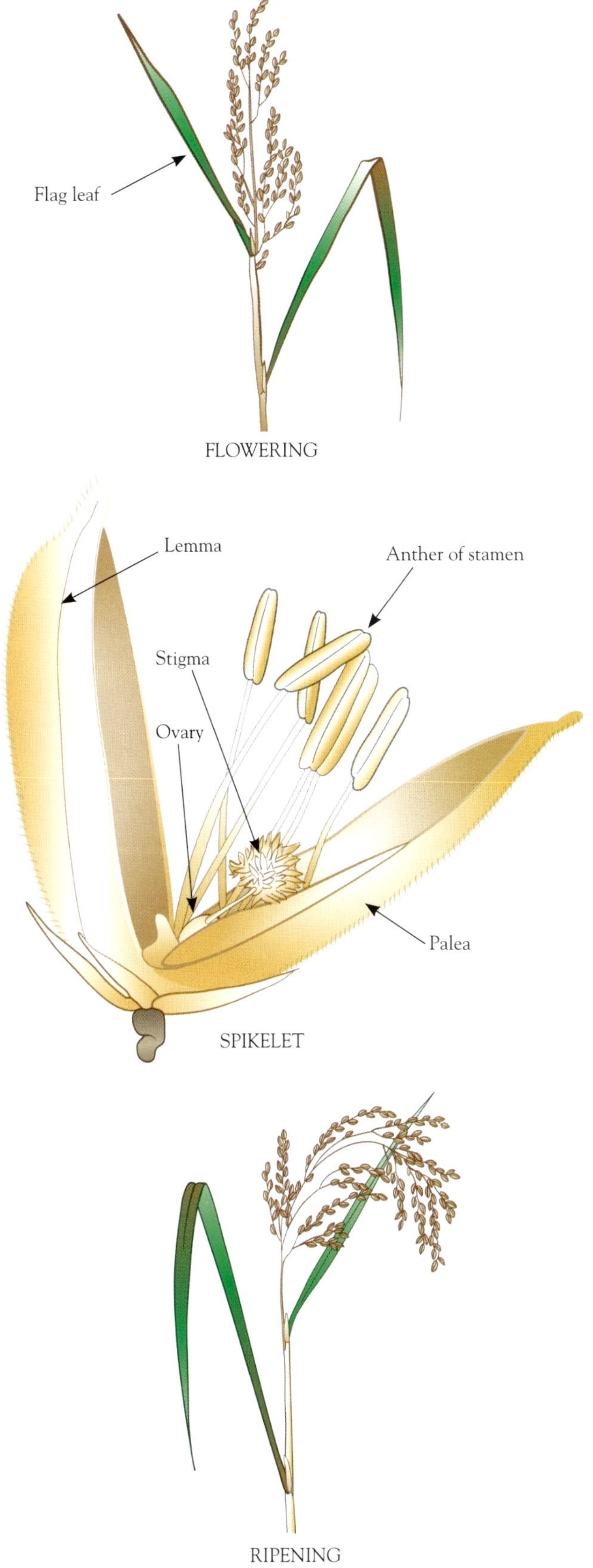

Figure 6. Flowering and ripening of the rice panicle. At heading, the panicle pushes through the flag leaf sheath (top). Once the whole panicle is out, it bends with the weight of the developing grain (bottom).

open are those closest to the main stem and at the panicle base. These are often sterile. Complete flowering in a single field may take 14 days or more, depending on variety, water depth and temperature, plant spacing, and supply of soil nutrients.

Anthesis (flower opening) normally occurs during morning or early afternoon hours (8 A.M. to 2 P.M.), depending on temperature. The hulls (lemma and palea) open, then pollen-bearing anthers are exerted and shed pollen onto the stigma. Rice flowers are nearly 100% self-pollinated. The lemma and palea then close, leaving the anthers outside. Fertilization occurs 5 to 6 hours later, by which time the percentage of spikelets that are fertile is determined. Each spikelet has one floret and potentially produces one grain.

The Ripening Phase

Grain Formation. After fertilization, the spikelets begin to build up starch to form grain. A rice grain is a ripened caryopsis—a single fertilized ovary surrounded by a hull made up of a lemma and palea (see Fig. 6). The developing kernel is filled with material stored in leaves and stems, which is remobilized and translocated to the grain, and from new carbohydrate produced by photosynthesis in the uppermost leaves. The majority of material used for grain filling comes from the flag leaf, so pest injury to the flag leaf has a larger effect on subsequent yield than injury to other leaves. During grain filling, the leaves begin to senesce, starting from the bottom of the plant. Grain filling and senescence of leaves are faster in some varieties than in others.

Duration of ripening depends on temperature and variety. Rate of ripening is faster in warm weather than in cool, and faster for long-grain types than for medium- or short-grain types. Varieties within a type also respond differently to temperature; for example, some long-grain varieties ripen faster than others. Grains formed during periods of high temperature are usually lighter and less well filled than grains formed during cooler weather, which reduces the quality of the harvested rice.

The grain goes through several phases during the ripening process, which may take 30 to 60 days. These changes are loosely termed water, milk, soft dough, hard dough, and ripe stages. Ripening is complete when the grain reaches its maximum dry weight.

Harvest Maturity. A rice kernel may be physiologically mature at a moisture content of about 27%. However, grain is harvested at a much lower water content to ensure the maximum number of mature kernels and to reduce the cost of drying the grain. Rice is mature and ready for harvest when 90 to 100% of the filled spikelets are hard and the hulls are yellow. It is ready for harvest when grain moisture content is reduced to approximately 22% or lower. Rice grown for seed production is harvested at a somewhat lower moisture level. Fields are ready for harvest about 45 to 50 days after 50% heading in normal years, depending on variety and location, and later in very cool years.

Growth Requirements

The basis of all rice plant growth is the capture of solar energy through photosynthesis, which occurs in the green parts of the plant that contain chlorophyll (Fig. 7). The sugar produced through photosynthesis is used for growth, maintenance, respiration, or compensation for damage caused by pests or environmental stress, such as excessively high temperatures. During grain filling, most of the carbohydrates produced during photosynthesis are translocated to the panicle, where they are stored in the grain. To effectively carry out photosynthesis, a plant needs water, light, oxygen and carbon dioxide, various nutrients, and sufficiently warm air and water temperatures; an insufficient supply of any of these requirements limits the plant's growth. Major objectives of an IPM program are to minimize the negative effects that pest injury has on photosynthesis and to use management practices that provide optimum conditions for photosynthesis, especially those that affect water temperature and depth and nutrient availability.

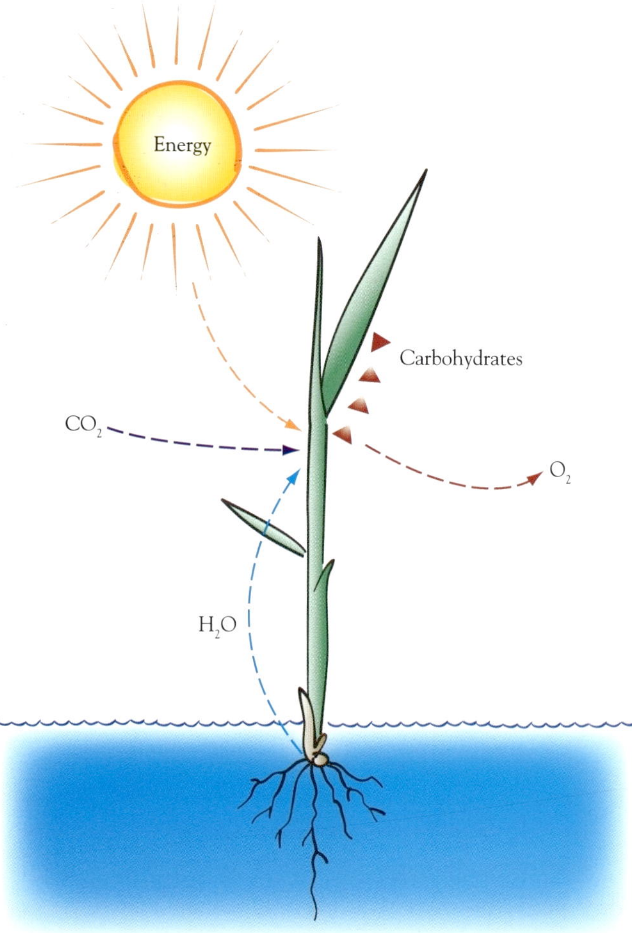

Figure 7. In photosynthesis, plants use energy from the sun to split hydrogen from water and combine it with carbon dioxide from the air to form carbohydrates. Oxygen from the split water molecules is released into the air. The carbohydrates (sugars and starches) are used immediately or stored for future plant growth and development.

Sunlight

The rice-producing areas of California have clear skies and high solar radiation during the growing season. Unlike tropical and subtropical areas, where cloudy skies may limit photosynthesis, solar radiation is rarely if ever a limiting factor for rice growth in California. However, too little light can be a problem in early stages of growth when weeds or muddy water obstruct the passage of sunlight to submerged seedlings. Cloudy or rainy weather during tillering and panicle initiation can reduce the number of panicles, spikelets, or filled grains; but these conditions rarely occur in California rice-growing areas during this critical developmental period.

Oxygen

Although rice is grown in standing water in California, it still requires oxygen. During germination, oxygen is necessary for biochemical processes responsible for metabolizing starch and converting it to plant tissue. There is sufficient oxygen at the soil-water interface to sustain germination and growth in flooded rice; however, oxygen concentration decreases with soil depth. Consequently, seed buried in soil by wind may not have enough oxygen for germination and growth.

During Seedling Stage I, the supply of oxygen affects the relative growth of roots and shoots. Rice growing underwater has less oxygen, resulting in more shoot growth and smaller roots; rice growing in unflooded fields has greater root expansion and shorter top growth. During Seedling Stage I or II, rice in deeper water has excessive long and weak top growth and underdeveloped roots, which can result in uprooting during windy periods.

At the start of Seedling Stage II, rice develops a system of specialized cells called aerenchyma, which conducts oxygen from leaves to roots. This is why it is important for the fourth leaf to be emerged from the water. Oxygen is not normally limiting to growth after the fourth leaf has emerged.

Temperature

Air and water temperatures are key factors in the development of rice and its pests. Establishment of the rice stand is favored by water temperatures between 77° and 84°F (25° to 29°C). Ideal air temperatures for rice production range from 77° to 90°F (25° to 33°C), except for the ripening stage when somewhat lower temperatures are better.

Prolonged low temperatures in the early growth stages reduce the development rate of seedlings. Slow-growing seedlings are vulnerable to seedling pests for a longer time and may suffer greater losses than those growing at optimal temperatures. Plants grown under low temperatures are shorter, have dark green leaves, and take longer to mature. Low temperatures inhibit the uptake of certain nutrients, notably phosphorus and zinc. Night air temperatures of 55°F (13°C) or lower can cause spikelet sterility during the 2-week period occurring 21 to 7 days before heading, when pollen is developing (see photo on p. 7). Raising the water depth to 8 inches (20 cm) or more after panicle initiation helps to protect the developing panicle from excessive cold. Some varieties of rice are more tolerant to cool weather, and these should be used in areas where cold weather is a problem.

After flowering, dry winds and temperatures above 95°F (35°C) or low temperatures around 50°F (10°C) can adversely affect pollination and seed set. Excessively cool temperatures during ripening can result in irregular maturity and incomplete ripening. If fields are drained too early during ripening, high temperatures and water stress may cause incomplete formation of kernels. Dew or rain on mature dry kernels may cause sun-checking (cracking or fissuring) from moisture reabsorption. Periods of north wind do not directly cause kernel fissuring; fissuring occurs after the wind stops and dew returns, rehydrating the dry kernels. Kernels with fissures or cracks are susceptible to breakage during harvesting and milling operations. Refer to the *Rice Quality Handbook*, listed in the suggested reading, for a detailed discussion of the environmental conditions that contribute to kernel fissuring and loss of milling quality.

Water

In California, most rice fields are flooded continuously from a few days before seeding until a few weeks before harvest. Flooding minimizes soil temperature fluctuations, suppresses growth of some weeds, enhances the availability of plant nutrients, and promotes high yields. Because of the rice plant's semiaquatic nature, rice grown under continuous flood consistently outperforms rice growing under nonflooded, upland conditions, yielding more kernels per panicle and achieving maturity earlier. Flooding slows seedling development, but within 30 days plants are taller and have longer leaves and more tillers than nonflooded rice. Rice grown in California uses an average of 3.5 acre-feet (4,317 cu m) of water through evaporation and transpiration (evapotranspiration). This is comparable to the amount of water used by alfalfa, irrigated pasture, and some deciduous orchard crops. Water movement out of the field by percolation and drainage adds to the total amount of water needed to grow a successful rice crop. Percolation and drainage are not considered to be actual water use since this water remains in the hydrologic basin.

Irrigation water quality is usually not a problem in California. Most of the rice acreage is irrigated with surface water originating from snow melt, and snow melt water is comparatively low in dissolved salts. In some areas, particularly in the Sacramento Valley, chlorosis caused by zinc or iron deficiency occurs in seedling rice plants when well water or drainage water that is high in bicarbonate is used for irrigation.

Low water temperature in many areas may delay development and reduce yield of rice in parts of the field. The yield loss associated with low water temperatures may be extensive in intake checks.

Nutrients

Rice plants require an adequate supply of mineral nutrients to carry out normal functions. Nutrient deficiencies retard growth, and deficient plants display symptoms that may resemble pest damage. Descriptions and photos of deficiency symptoms can be found under "Abiotic Disorders" in the chapter "Diseases." Nutrient-deficient plants have weaker photosynthesis and less growth than normal plants; they compete poorly with weeds and are less able to tolerate insect and disease damage. Over-fertilization with nitrogen may cause excessive vegetative growth, increased stem rot, blanking, lodging, increased weed growth, and delayed crop maturity. Using fertilizers properly, as described later in this manual, can prevent and sometimes correct these problems.

Lodged rice.

Photo: Mike Poe

Managing Pests in Rice

Integrated pest management treats pests as part of the total rice production system, which includes not only the crop and its pests but also the physical and biological environment in which the crop is grown. The goal of an IPM program is to coordinate pest management activities with all production practices that impact pests to achieve economical and long-lasting solutions. The emphasis is on anticipating and preventing problems whenever possible.

Four components are essential to any integrated pest management program:

- pest identification
- field monitoring
- control action guidelines
- effective methods for prevention and control

Each field needs a slightly different set of production and pest management methods for greatest profitability, and each field's requirements change from year to year. Climate, soil type, crop and pest history, cultural practices, the cultivar grown, and the nature of the surrounding land (topography and vegetation) all affect pest problems. The general discussion of management practices that follows should help you design a program tailored to the specific needs of your field. Table 1 lists the key management considerations to be made at various stages of crop development.

Pest Identification

The mere presence of invertebrates does not necessarily mean that they are causing economic damage. Many species are beneficial or innocuous; others cause damage only during certain stages of crop development. Before attempting to treat a pest problem, be sure that damage symptoms are those typically associated with the pest. Also, certain nutrient deficiencies or chemical toxicities can cause symptoms similar to those caused by pathogens or invertebrates, so these should be carefully diagnosed before taking a corrective action.

Most pest management tools, including pesticides, are effective against only a certain range of pest species, so accurate identification of the damaging organism is essential. Different methods or pesticides may be needed even for closely related species. The descriptions and photographs in this manual will help you recognize pests commonly found in California rice fields. Check other references cited in the text for further information. Remember that some pest problems can be diagnosed reliably only by experienced professionals; don't hesitate to seek technical help if you are not sure of an identification.

Table 1. Timetable of Key Management Considerations during Rice Development.

Growth stage and management considerations	See page
Seed Soaking, Draining, and Planting Stage	
seed quality	16–17
overheating, oxygen deprivation	5, 16–17
seed treatment for bakanae	16, 67*
delay of planting by bad weather	16
prepare for mosquito management	60–62
Seedling Stage 1 (germination to 3-leaf stage)	
seed rot and seedling diseases	68*
tadpole shrimp	47–48*
midge larvae	49–50*
crayfish	50–51*
monitoring rice water weevil	52–55*
waterfowl, blackbirds, and rats	83–88
water too cold or too deep	17
field drained, water too shallow	15, 17
seed burial or drift, bunched plants	16–17
zinc or iron deficiency	10, 18, 77–79
weed seedling identification	28–43†
monitoring weeds for possible resistance	26–27
water-holding requirements for pesticides	17*
herbicide injury	21, 81–82
Seedling Stage 2 (4-to 5-leaf stage)	
tadpole shrimp	47–48*
crayfish damage to plants and irrigation system	50–51*
bakanae	73–74*
rice blast	74–76*
rats and waterfowl	83–88
plant nutrient deficiencies	10, 18, 77–79
weeds and algae	28–43
water-holding requirements for pesticides	17*
herbicide injury	21, 81–82
Tillering	
nitrogen, phosphorus, potassium, and zinc deficiencies	10, 18, 77–79
rice water weevil damage	54–55*
weeds	28–43
herbicide injury	21, 81–82
stem rot and aggregate sheath spot	69–73*
rice blast	74–76*
gas and organic acid toxicities	80
Maximum Tillers	
nitrogen and phosphorus deficiencies	10, 18, 77–79
late weed growth	28–43
herbicide injury	21, 81–82
stem rot and aggregate sheath spot	69–73*
rice blast	74–76*
rice water weevil damage	54–55*

Growth stage and management considerations	See page
Panicle Initiation	
pollen damaged by cool weather; raising water depth	7
armyworms	57–59*
rice water weevil damage	54–55*
leafhoppers	56*
stem rot and aggregate sheath spot	69–73*
rice blast	74–76*
hydrogen sulfide toxicity	80–81
Jointing	
excessive vegetative growth from overfertilization	10
armyworms	57–59*
leafhoppers	56*
pollen damaged by cool weather; raising water depth	7, 9
crayfish damage to irrigation system	50–51*
muskrat damage to irrigation system	88
late weeds, monitor for possible resistance	26–43
intermittent drought from inadequate water supply	9
rice blast	74–76*
hydrogen sulfide toxicity	80–81
Heading	
blanking and lodging from excessive nitrogen	7, 10
blanking caused by overly cool or hot weather	6–9
waterfowl	83–85
armyworms	57–59*
panicles damaged by heat or wind	9
rice blast	74–76*
false smut	77
kernel smut	76–77*
hydrogen sulfide toxicity	80–81
Grain Formation	
lodging due to excessive nitrogen or stem rot	10, 18, 69–71
stem rot	69–71*
rice blast	74–76*
false smut	77
kernel smut	76–77*
blackbirds	85
rats, muskrats	85–88
waterfowl	83–85
draining too early	9
Maturity to Harvest	
maturation slowed by overly cool weather	9
proper kernel moisture content at harvest	8, 18
preparation and sanitation of on-farm drying and storage facilities	19
premature or late field drainage	9
waterfowl, rats	83–88
Postharvest to Preplant	
stored grain pests	19, 85, 87–88
crop residue management	18–19
precision leveling	15–16
seedbed preparation for next year	16
rat and muskrat control	85–88

*See UC IPM Pest Management Guidelines: Rice, listed in suggested reading (online at www.ipm.ucdavis.edu).

†See UC IPM Weed Photo Gallery (online at www.ipm.ucdavis.edu).

Your farm advisor, pest control adviser (PCA), or agricultural commissioner's office can provide assistance or direct you to professional diagnostic services.

Exotic Pests. Exotic species are plants, animals, or microorganisms that are not native to a particular region. Some exotic species have been intentionally introduced to California, while others are introduced accidentally. Some rapidly colonize an area and become serious pests, often because they are not controlled by natural enemies that limit their numbers in their native habitats. These rapid colonizers are often called exotic invasive species. Exotic species that are of concern in California rice production include red rice, monochoria, and panicle rice mite. Growers and PCAs should be on the lookout for these and other unusual species that might be exotic invasives. Once established, invasive species are extremely difficult to eradicate and can cause substantial economic and ecological problems. If you find suspected invasive species, contact your farm advisor and county agricultural commissioner's office.

Field Monitoring

Field monitoring provides information on daily or seasonal field conditions that can be used to predict and evaluate potential pest and nutrient problems. Because conditions vary even between neighboring basins, every basin within a field should be monitored. Make regular checks of the pest species present, the growth stage and health of the crop, weather, environmental conditions, and when appropriate, the population levels of pests and beneficial organisms. Keep records of monitoring results, weather, and management activities. Simple tables and graphs of data help define patterns, and maps help identify localized problems and pest movement. Use this information, together with records of control measures used, dates when cultural practices were carried out, and previous crops planted in the field, to help decide the need for future control measures. *UC IPM Pest Management Guidelines: Rice*, listed in the suggested reading, describes methods for monitoring each pest and includes sampling forms for keeping records.

Weather is a major factor controlling the development of the rice plant, its pests, and the natural enemies of those pests. Good weather information is necessary for timing cultural practices and for predicting pest outbreaks. Many public news outlets on radio, television, and the World Wide Web report local weather. The National Weather Service broadcasts local and regional weather on NOAA Weather Radio (VHF channels 162.40 to 162.55 megahertz) and on the Web (www.nws.noaa.gov). Rainfall, temperature, and evapotranspiration information for irrigation management is available from the California Department of Water Resources' CIMIS program (www.cimis.water.ca.gov).

For the most locally accurate data, set up a weather station in or near your fields. Available instruments vary from simple maximum-minimum thermometers to electronic devices that continuously monitor and record weather information for transfer to a computer. More sophisticated stations transmit the data to a remote computer. Set up and maintain the weather instruments according to manufacturers' instructions. Calibrate the instruments regularly to ensure accuracy. Compare weather information to your other monitoring records to help plan your crop and pest management programs.

Control Action Guidelines

Control action guidelines indicate when management actions, especially pesticide applications, are needed to avoid eventual yield losses from pests. Guidelines for invertebrate pests are often expressed as numerical thresholds indicating the level of infestation or the amount of damage that will cause economic loss. Guidelines for weeds, diseases, and vertebrates are usually based on the history of a field or region, the stage of crop development, weather, pest distribution, and other field observations. Control action guidelines are helpful only when used together with accurate pest identification and careful field monitoring. Specific guidelines for each pest are given in *UC IPM Pest Management Guidelines: Rice*, listed in the suggested reading and available online at the UC IPM website, www.ipm.ucdavis.edu.

Management Methods

The preferred methods in an IPM program are those that protect the crop while interfering as little as possible with the long-term sustainability of the production system. The cheapest and most reliable way to deal with pest problems is to anticipate and avoid them. When pesticides are needed, choose materials and application methods that control pests effectively with a minimum of side effects.

The most critical management decisions are usually made during the first 4 weeks after planting the crop. Often the decisions that must be made involve trade-offs; for instance, using deep water for grass control may slow rice development. Use of certain insecticides may delay or prevent application of some herbicides or require that water be held in the field for several weeks. Thus, when making one decision, you must try to anticipate other problems that may arise in the next few weeks.

Weather is often unpredictable during this time of the year, further complicating the situation. Wind may be extremely disruptive, affecting stand establishment, water depth and turbidity, and your opportunity to use pesticides safely. Cool temperatures delay rice growth and prolong the crop's period of vulnerability to many seedling pests. Frequent field checks, good planning, experience, and often a little luck with the weather are necessary for the most successful integrated pest management program. Decisions and scheduling of operations vary from year to year.

More detailed and current information on the crop management practices discussed in this chapter can be found in the

California Rice Production Workshop Workbook, listed in the suggested reading, and at the University of California Rice Research and Information Program website (www.plantsciences.ucdavis.edu/uccerice/).

Cultivar Selection

Many rice cultivars (varieties) adapted to California's growing conditions are currently available. Cultivars differ in type of grain, yield, stature, developmental time (time to maturity), tolerance to herbicides, blanking resistance, disease resistance, and other characteristics. Rice cultivars are usually loosely classified according to the time from seeding to maturity. Early and very early cultivars are comparatively insensitive to day length while most intermediate and late cultivars require a shortening of day length to initiate flowering. (See Figure 2, in the chapter "The Rice Plant: Development and Growth Requirements," for approximate developmental times for early and late cultivars.) New cultivars may be released and old ones withdrawn each year, so check the *California Rice Production Workshop Workbook*, listed in the suggested reading, and the University of California Rice Research and Information Program website (www.plantsciences.ucdavis.edu/uccerice/) for the latest information. Recently, cultivars have been developed for resistance to rice blast, but no cultivars are available that have been specifically bred for resistance to invertebrates in California. There is an ongoing program to develop cultivars resistant to rice water weevil, as well as the diseases aggregate sheath spot and stem rot.

Marketing considerations, yield, seed availability, time of planting, maturity date, water depth, and harvesting capacity are important factors involved in selecting a cultivar. Remember that your choice of cultivar indirectly affects pest problems because cultivars that are poorly adapted to local conditions grow less vigorously and are more susceptible to damage. If you are located in a district that has cold water or a history of cool early seasons, plant cold-tolerant cultivars. Always choose cultivars that mature in your locality sometime between mid-September and mid-October to avoid hot summer heat at grain maturation or early rains that can complicate harvest or damage ripening rice. Your farm advisor can provide you with information listing the latest cultivars and their characteristics.

The availability of high-yielding semidwarf cultivars has encouraged a management-intensive approach to rice farming. This approach includes higher rates of nitrogen and phosphorus fertilizers, higher seeding rates, precision laser-guided land leveling, and early-season shallow water management. These practices have increased yields and reduced per-unit production costs, but have also aggravated problems with weed competition, rat and bird depredation, and stem rot and aggregate sheath spot diseases. Minimize these problems by

- basing nitrogen needs on leaf tissue analysis
- avoiding overly dense rice stands
- maintaining a water depth (4–5 in, or 10–12 cm) that limits weed, bird, and rodent problems
- using herbicides carefully to control weeds

Cultural Practices

Many cultural practices can be manipulated to minimize pest damage. Making these adjustments is often the most economical and reliable long-term defense against pests. Land leveling and field design, seedbed preparation, planting rates and dates, water management, fertilization, harvest practices, postharvest crop residue management, general sanitation practices, and fallowing or crop rotation all have a significant impact on the damage caused by pests in rice. Rotating to rice can also be an effective way to manage pests in other crops if soil conditions permit. In addition, cultural practices can limit mosquito breeding in rice fields.

Site Selection. Rice production requires soils with low infiltration rates to prevent excessive water percolation and to allow water to warm. Preferred soils have a high clay content (35–50%) in the topsoil or subsoil, or a hardpan in the subsoil. Fields with deeper topsoils are the most productive, although good rice yields can be achieved on shallower soils if proper nutrition is provided. The topsoils of fields in naturally flat locations are less disturbed during field leveling than are fields on steeper land, where less fertile subsoils are exposed during leveling. Fields where a calcareous (high-calcium) or sodic (high-sodium) subsoil is exposed during leveling are particularly difficult for rice production because they often have soil chemistry problems that are difficult to correct.

Precision Leveling. Accurate leveling can reduce pest problems by improving the uniformity of the slope in basins, reducing the number of levees, and increasing basin size.

A poorly leveled basin (the area between levees, also called a paddy or a check) makes it difficult to maintain a reasonably uniform water depth: some areas within a basin may be too deep or too shallow. In deep water areas (7–8 in, or 18–20 cm), leaves of tall-statured rice cultivars may lie on the water surface and be more susceptible to rice leafminer attack. Shallow areas with little or no floodwater provide ideal spots for weed infestations and vertebrate pests to feed on rice seed or seedlings. In fields continuously planted to rice, basins may be leveled with a slope or fall of 0 to 0.05 feet per 100 feet (0–0.02 m per 30.5 m), whereas in areas where rice is rotated with row crops, a slope of 0.1 or 0.2 feet per 100 feet (0.03 or 0.06 m per 30.5 m) is commonly used. The flatter slope helps to maintain a uniform water depth at the recommended 4- to 5-inch (10.2–12.7 cm) level early in the season and thus avoid these problems.

To minimize weed and weevil problems, levees should be made parallel, reduced in number or completely eliminated, and basin size increased. Levees offer sites for weeds to grow and produce seed. Barnyardgrass, which frequently grows on levees, may scatter seed and become a problem in the field if water is drained or too shallow. Weedy levees also attract rice water weevils in spring and subsequently provide them with overwintering sites. Fields with many levees and small basins have relatively greater weevil infestations than those with larger basins and fewer levees. Fields with straight levees, unlike those with narrow contour levees, permit water weevil treatments along the levees and field borders, thus minimizing the amount of pesticide applied.

Generally, make basins as large as practicable, but be aware that the larger the basin, the more it is subject to wind and wave damage. Large fields tend to flood slowly and encourage midges, tadpole shrimp, and weed infestations to get started ahead of the rice. This frequently results in greater pest damage and can be corrected by splitting the field into two smaller fields when leveling, or by modifying the irrigation system to flood faster.

Tillage and Seedbed Preparation. The way you prepare your field before planting can have an impact on your losses to pests and other problems. For weeds, tillage in fall or late winter enhances germination of weed seeds in the surface soil. Sprouting weeds can then be destroyed with spring cultivation. If you have severe weed problems, consider plowing 6 to 8 inches (15 to 20 cm) deep with your first cultivation in fall or early spring; this practice buries some weed seeds deep enough to prevent them from emerging. Subsequent tillage operations should be shallow (2–3 in, or 5–7 cm) to avoid bringing buried seeds to the soil surface, where they can germinate along with the newly planted rice. Drill-seeding rice without spring tillage reduces weed problems because weed seeds are not brought to the surface. For diseases, tillage in fall hastens the decomposition of crop residues that harbor stem rot and other diseases. Deep plowing can also bury stem rot sclerotia and, if subsequent tillage operations are shallow, keep a portion of them from floating and infecting rice plants at the waterline. Thorough incorporation of rice straw residues in fall also reduces algae and the generation of toxic gases and organic acids that can injure rice plants in the spring and summer.

Land planing or smoothing the soil with 3- or 4-wheel planes reduces soil clod size and other irregularities on the leveled seedbed surface. High spots or large soil clods, when exposed above the water surface, often provide sites for growth and survival of weeds. As the final tillage operation, groove the soil with a corrugated rice roller across the prevailing wind direction to provide a place for young seedlings to anchor and to reduce uprooting of seedlings by the wind or waves. The finished seedbed should be dry to at least the 4-inch (10 cm) depth and well aerated to help minimize anaerobic soil conditions upon flooding, which slow seedling growth.

Seedbeds that will be dry-seeded are prepared in the same way as for water-seeding except that the finished seedbed is worked more thoroughly to be finer and well packed so seed depth can be controlled precisely. Alternatively, rice may be drill-seeded without tillage if the soil surface is not rutted from the previous harvest. Such fields may be tilled in the fall to leave an even surface for spring planting. Such no-till planting reduces tillage costs, gives the rice crop an earlier start, and discourages weeds, which tend to be less severe when seeds are not brought to the surface by tillage. Heavier, specialized drills are needed to plant through crop residue and packed soil.

Borrow pits are deep channels from which soil has been removed to form levees. They provide a site for weeds and may be graded out where permanent levees are used. In fields that are rotated to row crops, removing borrow pits is probably not justified by the expense and effort required. Fields with borrow pits may be easier to drain, and these deep-water areas provide temporary protection for mosquito fish if fields must be drained for short periods of time. Removing them could make mosquito control more difficult.

Planting. The ideal time for planting depends on cultivar, field conditions, and weather, but typically it is after temperatures warm in the spring. Planting dates range from mid-April to late May. Planting too early can result in stand losses from cold weather. Later planting can reduce yields. Use certified seed; it reduces losses from low germination or weak seedling growth and guards against cultivar mixtures and weed seeds.

When water temperatures fall below 70°F (21°C), seed and seedlings grow very slowly and, as a result, suffer more damage from seedling diseases and invertebrate pests, compete poorly with weeds, and may suffer zinc deficiencies. Prolonged exposure to water temperatures below 65°F (18°C) will reduce yield. Seed treatment can protect against seedling diseases and bakanae, and treated seed should always be used when it is necessary to plant under cool conditions. Consult the *UC IPM Pest Management Guidelines: Rice* for information on seed treatment materials and how to use them. In addition to seed treatment, seed that will be planted into flooded fields should be soaked for 18 to 24 hours and drained for 24 to 36 hours to induce the germination process.

It is beneficial to plant immediately after flooding to optimize soil oxygen and to minimize losses from midges, tadpole shrimp, and weed competition. Immediate planting also helps avoid slick soil conditions that can contribute to seed drift. Seed should be soaked during the last 40 to 60 hours of flooding to ensure you will be able to plant as soon as possible after flooding is completed. If planting is delayed, flush drained seeds with cold water or aerate them with cool air. Heat produced by germination in the soak tank can be lethal to seed if temperatures remain above 105°F (40°C) for 6 to 12 hours. Where fields flood slowly, start seeding the upper end of the field before water reaches the lower end.

Choosing the proper seeding rate is important in limiting losses to several pests and when using numerical guidelines for treatment of some invertebrate pests. A seeding rate of 150 pounds per acre is suggested throughout this manual. Using more than 150 pounds per acre (168 kg per ha) will produce dense plant stands, which can increase stem rot and result in weak straw that increases lodging, ultimately reducing yield. If all seed planted at this rate were to successfully germinate and establish, the stand density would be about 40 to 60 plants per square foot. However, losses to invertebrates, vertebrates, seedling diseases, seed drift or burial by wind and wave action, and environmental stresses may reduce this number by one-third to one-half during the first few weeks of stand establishment. Using the suggested rate will result in an initial stand density of 10 to 20 seedlings per square foot, the density that is needed for uniform grain ripening and acceptable yield, as well as a shading out and suppression of weeds.

Water Management. Good water management is the key to preventing many pest problems. Fields should be flooded as rapidly as possible, preferably within 2 to 4 days of final seedbed preparations, and then planted without delay. Rapid flooding and planting help avoid or minimize problems with midges, tadpole shrimp, and weeds. Rapid flooding also helps maximize soil oxygen levels at planting to encourage rapid stand establishment. Some fields may require changes, such as increasing the size and/or number of inlet and weir (rice box) structures, to flood more quickly. Some large fields may also need to be split into smaller ones to encourage rapid flooding. In all cases, flood only as many fields at a time as you have adequate water for. Thirty gallons per minute per acre is an adequate rate for flooding a field; this rate fills a typical 100-acre field of clay soil in about 4 days. To maintain an adequate flow rate, promptly repair all irrigation system leaks.

Plant rice in well-leveled basins with an average water depth of 4 to 5 inches (10–13 cm); this water depth offers a good balance between vigorous stand establishment and suppression of several pests. Water that is too shallow (less than 2 in, or 5 cm) or uneven in depth may expose soil, promoting weed growth and allowing access for rats and birds. Water that is too shallow may also slow rice growth when the weather is cool. On the other hand, deeper water (5–7 in, or 13–18 cm), while it suppresses several important weed species and reduces rat and blackbird problems, slows early-season rice growth. Rice grown in deeper water emerges through the water later, is taller and weaker than shallow water rice, and has delayed leaf and tiller development. Seedling leaves may float on the surface of deep water instead of being upright and are therefore more vulnerable to leafminer damage. Most early-season delays in rice growth are temporary and are usually not evident by midseason. By late season, leaf development is often more advanced, heading is earlier, and grain reaches the harvest moisture content sooner in rice grown in deeper water than in rice grown in shallow water.

Establish and adjust water levels in fields as needed to discourage pests and promote rice development. For most situations a moderate water depth (4–5 in, or 10–13 cm) is best. If a field has watergrass problems, deep water (7 in, or 18 cm) can be used in combination with standard herbicides to improve control. Regardless of the early-season depth, slowly raise or maintain the depth to at least 8 inches (20 cm) at panicle initiation (55 to 60 days after planting) to insulate the developing panicles from cold air temperatures and help ensure the formation of viable pollen. This water level can be gradually lowered after heading.

When applying pesticides to fields, be sure that the irrigation system is in good order, that drain outlets are boarded and blocked with plastic and soil, and that tailwater is held for the time required on the label or permit. Also, before postflood applications are made, be sure that the desired water depth has been established, that all weir structures are boarded, and that water inflow is shut off. If water will be held for several weeks or if the permit requires it, the weirs can be blocked with plastic and soil. Holding water static in the field improves the effectiveness of some pesticides, keeps water from getting too deep in the bottom basins, and keeps pesticide residues out of public waterways. Water may need to be added to the field to maintain an adequate depth. However, some herbicides require that water be held static for a certain period of time. Do not allow the water to get too shallow or soil to be exposed as this may promote weed growth and, if it is cool, retard rice growth.

In most cases, fields should remain continuously flooded from just before planting until they are drained for harvest. Drain your field if necessary to apply foliar herbicide, manage toxic gases and acids, or control midges, tadpole shrimp, or crayfish. Draining the field may promote weed growth, increase nitrogen losses, and stress the rice plants, depending on how long the field remains drained, so take preventive measures against these pests to avoid draining whenever possible. When draining is essential, drain and reflood as fast as possible and follow label directions and water-holding requirements for pesticides already applied in the field. Remember that chemicals applied into the water are most effective when they are held in the field without draining. Well-graded fields drain and fill faster, but if the slope is too steep, one end of the field is likely to have too much or too little water, especially during stand establishment when water depth is critical.

Draining fields prematurely before harvest causes symptoms of moisture stress, including shriveled kernels and unevenly maturing plants that result in poor yields of rough and whole kernel milling rice. The best time to drain depends on soil type, cultivar, drainage facilities, and weather. On typical clay-textured rice soils, drain the field about 25 days after the rice is fully headed. Another guide is to drain when 70% of the panicles have tipped kernels in the hard dough stage, which corresponds to roughly 3 weeks before the projected harvest date.

More information on water management can be found in the *California Rice Production Workshop Workbook*, listed in the suggested reading.

Fertilizing. Maintaining adequate levels of nutrients and avoiding excess levels of nitrogen help ensure good yields of high-quality rice and reduce problems with weeds and diseases. Sample the soil for phosphorus, potassium, and zinc at least once a year in problem areas and every 2 to 3 years in fields where a previous soil analysis showed high nutrient levels. Take the samples well before planting, allowing enough time to analyze them and to apply necessary supplements before flooding. Phosphorus, potassium, and zinc are most effective when applied before flooding. Most fields are deficient in nitrogen and require annual applications of this nutrient. Detailed information on analyzing soil and plant tissue for nutrients and recommended nutrient levels can be found in the *California Rice Production Workshop Workbook* and the *Rice Quality Handbook*, listed in the suggested reading.

In the case of nitrogen, apply an adequate amount to optimize yield, but avoid excessive rates to minimize lodging, blanking, delayed maturity, and the potential for greater weed and stem rot problems. Split applications may be used to avoid high nitrogen levels early in the season or where rice is drill-seeded and flush-irrigated for stand establishment. Fields that were fallow or planted to green manure or row crops the previous year usually require less nitrogen than fields planted to rice or wheat. In addition, fill areas in recently leveled fields frequently require less nitrogen than cut areas. Check the references mentioned above for more details.

Some nutrient deficiencies and toxicities cause rice plants to develop distinctive symptoms. These symptoms are described and pictured in the chapter "Diseases." Always confirm suspected nutrient deficiencies or toxicities with a soil or leaf analysis before taking corrective measures. Other disorders can produce similar symptoms.

Harvesting. For maximum yield of high-quality rice, it is important to harvest when moisture content of the grain is at the proper level. Detailed information on recommended harvest moisture levels and proper harvest procedures can be found in the *Rice Quality Handbook*, listed in the suggested reading.

Postharvest Straw Management. Crop residue management is important in limiting stem rot and aggregate sheath spot diseases, toxic organic acids and gases, weeds, algae, and various other problems. Straw residue management options include incorporation into the soil, complete removal, or burning. Although burning is currently the most effective and economical method, growers are allowed to burn only up to 25% of their rice acreage, but because of air quality considerations burn only about 13% of the acreage in a given year. In some cases, straw may be left standing until spring without incorporation or winter flooding. This is not a recommended practice but may be done where there is a reasonable certainty the field can be burned in the spring.

Residue Incorporation and Winter Flooding. Incorporating rice straw and then flooding fields and maintaining the flood through the winter months reduces disease problems as compared to incorporation without flooding. This method also increases the return of nutrients to the soil. The decomposition process requires good straw-soil contact and sufficient moisture, warmth, and time for soil microorganisms to decompose the residue. Chopping and spreading the straw before incorporating it, followed by winter flooding, helps satisfy these requirements and helps avoid problems with toxic gases and organic acids, algae, and nitrogen immobilization.

If you incorporate crop residues, chop and thoroughly incorporate the straw into the soil as soon after harvest as possible. Chopping straw into short pieces facilitates the movement of machinery and encourages more rapid and complete straw decomposition. Incorporating with a stubble disc or other implement before flooding improves decomposition but buries waterfowl feed. Delaying incorporation until spring, especially just before planting, aggravates straw-related problems.

Incorporate straw and flood fields as soon after harvest as possible. This allows more time for decomposition and, as an added benefit, encourages waterfowl activity. The initial water level should be 2 to 4 inches (5.1–10.2 cm) deep. If straw is not incorporated before flooding, specially built "cage rollers" may be used about 1 week after flooding to increase the contact of straw and soil and improve the rate of decomposition. After straw is incorporated, flood depth may be increased to 4 to 6 inches (10.2–15.2 cm) if water is available. Drain flooded fields by late February or early March to allow the soil to dry properly before fields are prepared for the next crop.

Incorporating rice straw and maintaining flooded conditions during winter returns substantial nitrogen and potassium to the soil and reduces the amount of nitrogen and potassium fertilizer needed for the next rice crop. The full benefit of nutrient return is not achieved until after 4 or 5 years of consistent incorporation and winter flooding.

Total Residue Removal. Cutting the straw at or near ground level (below the waterline) and then removing the residue as completely as possible is nearly as effective as burning in minimizing diseases and straw-related problems. The major drawback to total removal is the difficulty and expense of rice straw disposal since no significant commercial markets have been developed for it. Studies have shown, however, that it can be used as a livestock maintenance feed–especially if it is ammoniated—and as a mulch to prevent erosion, and to produce pressed boards, paper, and energy.

The goal of total residue removal is to collect and remove as much of the rice straw as possible. Stem rot infections and overwintering structures (in the form of sclerotia) occur principally at water level, so cutting rice at ground level removes most diseased plant tissue. Sclerotia that are loosened from straw residues or free in the soil do not compete well with soil microorganisms and lose viability rapidly during overwintering.

Removing straw is also effective in avoiding problems with organic acids and gases, algae, and nitrogen tie-up. If you decide to minimize stem rot and other straw-related problems by removing residues, follow these steps:

- Cut the straw at ground level below sites of stem rot infection; spread the straw or rake it into windrows.
- Bale, roll, or chop and remove the residue. Spread straw can usually be baled in 1 to 2 days after harvest; windrowed straw takes 4 to 5 days.
- Disc or plow remaining stubble in fall or spring.
- Long-term removal of residue may require fertilizer applications, particularly potassium, to replace nutrients removed with the straw.

Burning. A permit is required from the local air pollution control district before a field can be burned, and burning will be allowed only on specified days when weather conditions are favorable. Follow the steps below to prepare for your burn, and always carry out a thorough postburn tillage program to destroy any residues remaining after the burn.

- Harvest and spread enough straw to fuel the fire and completely destroy all residues; straw that has been spread dries faster than straw piled in rows.
- Limit vehicle traffic in the field at harvest time. Machinery pushes straw into the wet soil and slows drying.
- Mow your levees to favor spread of the fire and improve overall field sanitation.
- Check straw moisture. For good burning, moisture content should be below 12%. Rice straw is dry enough to burn if it crackles when it is bent sharply. If it rains, the drying process may have to be extended or repeated.
- Check all tools and equipment. Go over your burn plan with your crew. Failure to do this may result in injury to crew members.
- Check wind direction. Be sure that the smoke will not cause hazards in populated areas, along highways, or near airports.
- Make provisions for signs and flaggers to warn traffic on adjacent roadways.
- If permission to burn has been granted by the local air quality district and all the conditions above are met, light your fire using the into-the-wind strip lighting technique shown in Figure 8. This technique reduces the smoke particulates you send into the atmosphere by about 50% or more.

Storage. Most growers use local commercial or cooperative drying and storage facilities; but if you dry and store your own rice, take precautions to avoid infestations of storage pests and maintain safe moisture conditions (at or below 14% moisture) for storage. Remove all trash and weedy plant growth from the storage area. Make bins and buildings safe from rodents and birds, and treat them with an approved insecticide to protect grain from attacks by stored grain insects. Regularly inspect storage facilities to be sure they stay free of insect, rodent, and bird pests. Detailed information on handling and storage can be found in the *Rice Quality Handbook*, listed in the suggested reading.

Sanitation. Keep weed growth around your farming headquarters and on ditch and canal banks, levees, and road edges to a minimum. Use sheep, mowing, cultivation, and approved herbicides. Controlling weeds on levees and field edges reduces the attractiveness of fields to water weevil infestations, reduces habitat for rodents and exposes them to predators, and also reduces sources of weed seeds and propagules. Maintain roads so that you and your PCA can get to all parts of the farm for field monitoring and so that mosquito abatement district personnel can check for mosquitoes and mosquito-eating fish.

Figure 8. Start your fire by igniting the downwind edges of the field. Leave two 50-foot unlit entry spaces in the center of each edge about 300 to 600 feet (91–183 m) apart. Two lighters then walk from the entry spaces maintaining a line parallel to the wind. If the wind is shifting, follow the average wind direction. After the field is lit, stand by to make sure the burn is complete. When lighting an into-the-wind strip fire, walk into the wind slowly (at a rate of about 2 mph), looking backward to be sure the wind isn't changing. Always keep in constant eye or voice contact with your lighting partner using cell phones or walkie-talkies. Keep a wet gunny sack and a water pump handy to control the fire.

Fallowing and Crop Rotation. You can limit populations of certain pests by fallowing fields for 1 or more years or rotating to crops with life histories and associated cultural and chemical practices that are hostile to those pests. These practices are directed primarily against weeds, although certain diseases and invertebrate pests may also be significantly reduced.

Rotating to crops in which cultivation can be utilized during part of the growing season can be very effective in controlling nonaquatic weeds. Crop rotation also provides an opportunity to use herbicides that are highly effective against barnyardgrass and watergrass, although special care should be taken to avoid any herbicide that may be phytotoxic to subsequent rice plantings. Rotating to nonirrigated crops such as safflower can also help control crayfish and perennial weeds.

Rotations that include a fallow period of a year or more can provide some control of perennial weeds such as Gregg arrowhead, river bulrush, cattail, and American pondweed. Plowing 8 to 12 inches (20–30 cm) deep to expose rootstocks and other vegetative propagules to drying and freezing conditions also reduces the population.

Spring or summer flooding of fallow fields encourages the growth of annual and perennial weeds that can be subsequently controlled by cultivation. Fields should be flooded for about 2 weeks to germinate as many aquatic weeds as possible. Weeds must be destroyed before they set seed or develop other reproductive structures. The suitability of this practice depends largely on the cost and availability of water in your area.

Biological Control

Natural enemies, including predators, parasites, and pathogens, attack and kill certain pests in rice fields, occasionally exerting significant biological control of insect pests such as leafminers and armyworms. Natural enemies, especially introduced populations of the mosquito-eating fish, *Gambusia affinis*, can be very important in controlling mosquitoes in rice fields. However, for the most part, biological control is not presently as important in an integrated pest management program for rice as in other crops. Future research may lead to greater utilization of natural enemies against pathogens, weeds, and invertebrate pests in rice.

The mosquito predator fish *Gambusia affinis*, attacking a mosquito larva.

Pesticides

When properly used, pesticides can provide convenient, economical protection from pests that otherwise would cause significant losses. Pesticides are the only feasible means of control in many situations. However, careless or excessive use of pesticides can result in poor control, crop damage, pesticide resistance, and hazards to human health and the environment.

In an IPM program, pesticides are used only when field-collected information indicates that they are needed. For some pests, such as tadpole shrimp and armyworms, this means taking population samples and using the control action thresholds in the *UC IPM Pest Management Guidelines: Rice*. For other pests, such as rice water weevils, treatments are largely preventive, and field-monitoring records from previous years, together with information on the economics of the crop, must be considered. Eliminating unnecessary treatments can reduce costs and potential hazards.

Sometimes a pest can be treated with one of several chemicals; the choice depends on the degree of control necessary, the effect on other pests, natural enemies or wildlife, environmental safety, and economic and legal restrictions. Most pesticides are available in a number of formulations and can be applied at various rates and by various methods to obtain desired control safely and at an economical price. Before choosing a pesticide, check with your farm advisor and the *UC IPM Pest Management Guidelines: Rice*, listed in the suggested reading. Always read the pesticide label; it outlines safety precautions, legal requirements, and registered uses for each product. Some of the other problems that can arise with the use of pesticides are discussed below.

Pesticide Resistance. Pesticide-resistant strains of a pest are able to withstand the application of a pesticide at a rate that formerly killed most individuals of that species (Fig. 9). Resistance generally develops most quickly under the selective pressure of repeated pesticide applications. Weed resistance to herbicides has become a widespread problem in California. Some species of mosquitoes are known to be resistant to certain insecticides. Once resistance has developed, it becomes necessary to turn to other pesticides or to higher dosages to reduce pest populations below economically damaging levels, substantially increasing the costs of pest control. Alternative rice stand establishment methods may be used to manage resistant weeds. These are discussed in the *California Rice Production Workshop Workbook*, listed in the suggested reading. Relying on a number of different control methods and avoiding repeated applications of the same material or materials with the same mode of action slow the development of resistance. Management of herbicide-resistant weed populations, the most serious pesticide resistance problem faced by rice growers in California, is discussed in more detail in the next chapter.

Pest Resurgence and Secondary Outbreak. Pest resurgence and secondary pest outbreaks occur when pesticides kill or otherwise disrupt the activities of natural enemies or competitors of pests. When natural enemies are killing a substantial portion of a pest population, removing the enemies by pesticide application causes the pest populations to rise. When the affected pest is the one that the pesticide was applied to control, the phenomenon is called pest resurgence. When a pesticide application disrupts the natural control of a pest that was not previously a problem, it is called a secondary pest outbreak. Because of the limited use of insecticides in rice and the minor role of natural enemies in controlling pests in this crop, pest resurgence and secondary outbreak have not been important problems in California to date. However, copper compounds applied at high rates to control algae and certain insecticides sometimes kill the mosquito-feeding fish, *Gambusia*, causing a potential increase in mosquito populations.

Phytotoxicity. Phytotoxicity is injury to plants caused by toxic substances. Herbicides are the usual cause of phytotoxicity, but other pesticides, fertilizers, toxic substances produced when organic matter decays underwater, and air pollution can also injure plants. Two common phytotoxicity problems associated with rice are damage to rice from the improper application of herbicides and damage to neighboring orchards and crops due to herbicides drifting from rice fields.

Herbicides are formulated to selectively control weeds while being safe to rice, but under some conditions they may injure rice as well as weeds. They are successful only when applied properly and precisely. An application at the wrong rate, at the wrong time, with the wrong material, or under the wrong environmental conditions can kill or severely damage rice plants. Phytotoxicity can also occur when certain mixtures of herbicides and other chemicals are applied together or when applied soon after one another; for instance, propanil should not be mixed with malathion or carbaryl or applied within 14 days before or after an application of one of them. Rice is especially vulnerable to certain herbicides in the early seedling stages and after panicle initiation. Proper herbicide application is discussed in more detail in the *California Rice Production Workshop Workbook*, listed in the suggested reading.

Drift. Drift of aerially applied herbicides onto neighboring crops can be a problem. Severe restrictions have been placed on the aerial application of several herbicides. Use of ground application, larger droplets, improved nozzles, reorientation of nozzle direction, frequent calibration of nozzles, and strict consideration of atmospheric conditions can limit drift problems to some extent. Ground application is required for some materials. Methods for reducing herbicide drift are discussed in more detail in the weeds chapter. Check with your county agricultural commissioner for local regulations concerning use of herbicides.

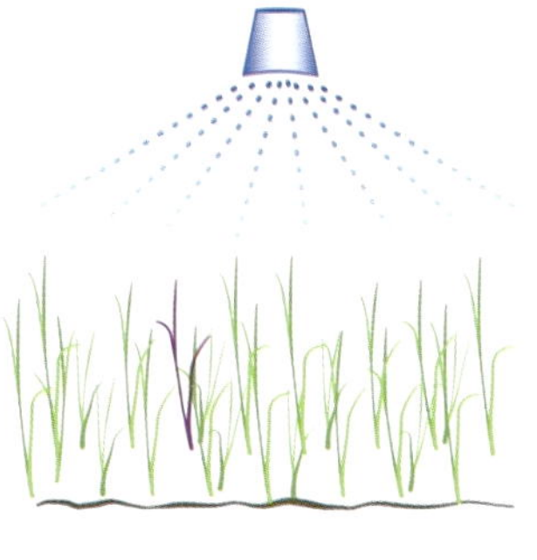

YEAR 1

There is always some finite probability that certain plants within a population are genetically resistant to the herbicide.

After applications

The only survivors, if the application is done correctly, will be the resistant plants, which will grow and set seed.

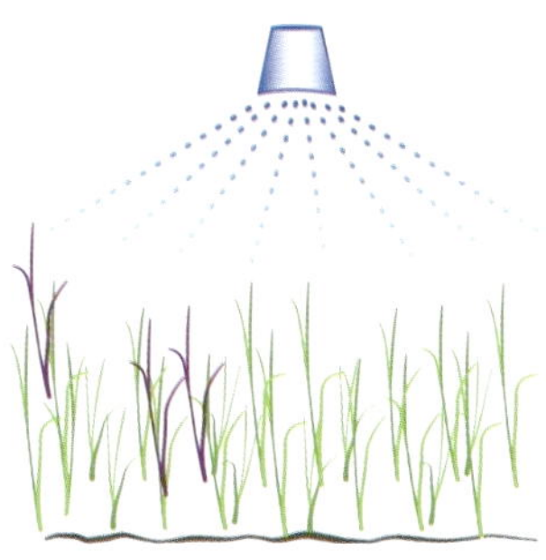

YEAR 2

Now there are more resistant individuals in the population. Application of the same herbicide or products with the same MOA will increase these individuals even more.

After applications

The remaining resistant population will then set seed.

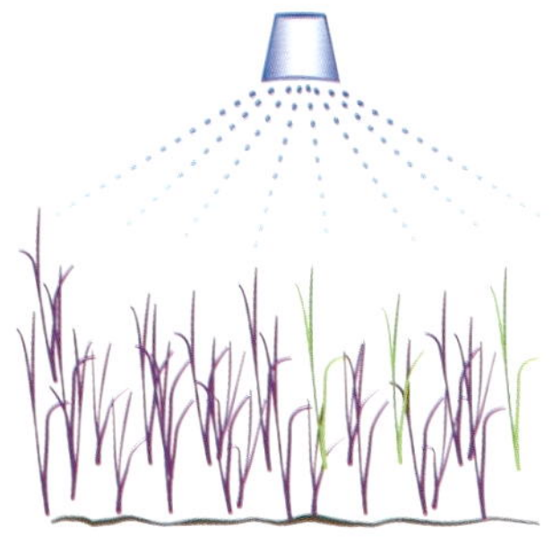

YEAR 3

Eventually, the population becomes mostly resistant individuals.

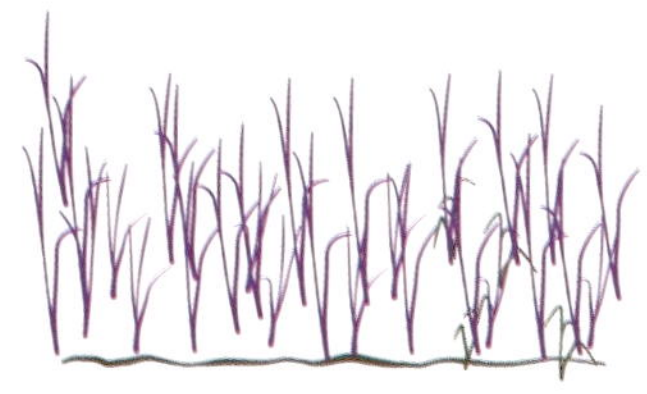

After applications

At this point, the herbicide is no longer effective.

Figure 9. Pest populations develop resistance to pesticides through genetic selection.

Pesticide Residues. Certain pesticides may remain in the water, soil, or on the crop for some time after an application. Follow all pesticide water-holding requirements specified on the label or use permit. These requirements allow sufficient time for pesticides to dissipate and break down in the field and help keep residues out of public waterways. Also follow other label restrictions to avoid the possibility of leaving illegal residues on stubble or grain. Before applying any pesticide, check the pesticide label for such use limitations. Before planting rice, be sure that you know what pesticides were applied in the previous crop; some may injure rice or have restrictions that do not allow a rotation with rice.

Hazards to Human Health. Some of the pesticides used in rice are hazardous to people. The people most at risk are loaders and applicators. Other people who spend time in the field such as fieldworkers and irrigators may also be exposed. People in surrounding areas may suffer pesticide poisoning when sprays drift from the field onto yards, roadways, and other public areas. Minimize health problems by following label directions and state and local regulations, wearing appropriate protective clothing, confirming the availability of emergency health care, and avoiding drift and unnecessary use of pesticides. See *The Safe and Effective Use of Pesticides*, listed in the suggested reading, for a detailed discussion on health hazards and safety precautions.

Hazards to Wildlife and Domestic Animals. It is also important to consider the hazardous effects pesticides may have on fish and wildlife and on domestic animals. Waterfowl can be killed if they are present in fields during or shortly after applications of various insecticides. Mosquito fish may need to be reintroduced after applications that might cause toxicity; local mosquito abatement districts work with growers to accomplish this. Fish or other aquatic life may be adversely affected or killed if pesticide-contaminated water from fields or water used to soak rice seed drains into public waterways. Follow label requirements, the requirements of your local county agricultural commissioner, and the standards of your local regional water quality control board to minimize these problems.

Hazards to Bees. Although bees are not normally found foraging in rice fields, they are sometimes killed by insecticide applications to rice when hives are placed in orchards adjacent to rice fields or when bees forage in flowering weeds found along levees or field borders. If any of these conditions are present around your fields, check for foraging bees before you apply insecticides. Many materials, especially those that might be applied to control armyworms, should not be applied when bees are foraging. Check with your agricultural commissioner for local regulations concerning bees.

Weeds

Weeds reduce yields by competing with rice plants for light, water, and nutrients. This competition is most critical from the seedling through the tillering stages. Weeds also lower the value of the crop by reducing quality, and they increase the costs of production, harvesting, drying, and milling. Weeds may contribute to insect problems by serving as alternative hosts for rice water weevils, armyworms, and leafhoppers.

Managing weeds has been a major concern for rice growers since the beginning of rice production in California. The California system of water-seeding rice into continuously flooded fields was developed in the 1920s as a cultural control for severe infestations of barnyardgrass. Although continuous flooding suppresses barnyardgrass, its use has encouraged other weed problems; the most serious of these is watergrass, a large-seeded relative of barnyardgrass. Most of the other major weeds are common native marsh plants; these plants, which include grasses, sedges, rushes, arrowheads, waterhyssop, waterplantain, pondweeds, and algae, thrive under continuous flooding and compete successfully with water-seeded rice.

Virtually all crop management practices in some way affect the competitive ability of rice and the weeds that infest rice fields. Soil fertility, seedbed preparation, time of planting, choice of cultivar, stand density, water management, pest damage, and previous crops all can have a dramatic influence on both the type and vigor of weeds in the rice field. For instance, invertebrates feeding on seedlings and young plants often slow the growth of the rice stand, lowering its ability to compete with weeds at the time they are beginning to establish themselves.

The introduction of short-stature rice cultivars into California and their widespread use present a good example of how modifications of crop management practices can interact to complicate weed control. Weeds have become increasingly difficult to control in fields planted with these cultivars; the reduced plant height and decreased shading ability of short-stature cultivars, coupled with the practice of shallow water for stand establishment, have created a better environment for weed germination and growth. Moreover, the higher rates of nitrogen fertilizer used to produce maximum yields from these new cultivars have added to weed as well as crop vigor. An integrated pest management program takes all these factors into account, together with a careful chemical and cultural control program, to optimize rice production. Neither chemicals nor cultural practices alone are adequate to give satisfactory weed control.

Prevention

The first step in a good weed control program is to prevent the introduction of weeds into the field. Always use good-quality rice seed and seed that is certified as weed-free. Certified rice seed has a maximum of 0.10% weed seeds by weight, with seeds of watergrass and barnyardgrass limited to less than 0.01%; certified seed has no red rice or noxious weed seeds. Irrigation water and farm machinery frequently transport weed seeds or other plant propagules into the field. Prevent introduction of weed seeds and vegetative reproductive structures such as tubers and rhizomes on machinery by cleaning combines, bankouts, and trucks when they move from fields with heavy weed infestations. Work fields with the fewest problem weeds, especially perennials, first.

To reduce field invasion, keep levees and roads as weed-free as possible and keep weeds in these areas from going to seed by using herbicides, mowing, or grazing. Levees often provide shelter for waterfowl and upland game birds, so plan your weed management program to preserve as much of their habitat as possible.

Cultural Management Methods

Certain minor adjustments in cultural operations can often make a big difference in the severity of weed infestations. These practices are not new to rice production in California, but they continue to be an important element in an integrated pest management program for weeds, augmenting and improving the effectiveness of weed control. For several weed species, they are the principal control method available in rice.

Tillage and Field Preparation. Field preparation can affect future weed problems in several ways. Tillage, land leveling, and preplant fertilizing are all important.

Tillage in fall or winter is suggested because it increases soil aeration and temperature and speeds drying of the soil, which helps desiccate vegetative weed propagules. Work rice seedbeds as soon after harvest as possible to hasten crop residue decomposition and decrease the incidence of algae problems next year. Tillage in the fall or winter enhances germination of weed seeds in the surface soil in early spring; then spring tillage destroys weed seedlings. However, in some years, especially in heavy clay soils, fall tillage may not be practical because the soil may be too wet. In fields that are flooded after straw is incorporated, using fall tillage to expose and desiccate the survival structures of perennial weeds is not possible. Fields heavily infested with perennial weeds should be specifically targeted for dry tillage in the fall rather than being flooded after straw incorporation. Straw incorporation by wet-rolling or tilling buries weed seeds, creating an overwintering seed bank that cannot be reduced by birds and small mammals.

Inadequate grading or planing of the field can increase weed problems. Water depth is easier to control in level fields, and level land requires fewer levees, thus providing fewer sites for weeds that thrive in areas where the soil is not adequately covered with water.

Grooving the rice seedbed with a heavy, ridged roller produces a uniform corrugated seedbed that will protect rice seedlings from wind and wave drift. Although heavy rollers may bring moisture to the surface of wet soils, increasing early weed establishment, rollers improve the uniformity of both rice and weed establishment and thus ease decisions for herbicide application timing based on crop or weed growth stage.

If a roller is not used, make sure the final seedbed is free of large clods. Clods so large that they are exposed above the water surface provide a favorable condition for growth and survival of grass weeds. Small clods about 1½ to 3 inches (3.8–7.6 cm) diameter are acceptable and help protect rice seed and seedlings from wind and wave action. Small clods also favor the establishment of a uniform stand, which is more resistant to weed infestations.

Fertilizer Management. Incorporate nitrogen and phosphate fertilizers into the seedbed at a soil depth of at least 2 to 4 inches (5–10 cm); this will reduce nitrogen losses and at the same time reduce the availability of both nutrients to weed seedlings germinating near the soil surface. For instance, submersed aquatic weeds such as southern naiad and algae grow more vigorously and often become well established when high rates of nitrogen and phosphorus are left on the soil surface. Applying phosphorus fertilizer 30 days after flooding, rather than before planting, greatly reduces populations of the blue-green alga *Nostoc*. Topdressing with nitrogen or phosphorus into the water before the rice canopy has covered the field surface also encourages rapid growth of weeds.

Water Management. Proper water management is the most important factor in controlling weeds in rice. Water management practices are greatly influenced by herbicides used in rice. The increase of weed resistance to herbicides applied into the water has resulted in the use of more foliar contact herbicides, which requires fields to be drained early in the season to ensure herbicide coverage of target weeds. These fields must be reflooded immediately to prevent germination and establishment of new weeds. Techniques for managing water with relation to herbicide applications are discussed in the *California Rice Production Workshop Workbook* and *UC IPM Pest Management Guidelines: Rice*, listed in the suggested reading.

Short-stature rice cultivars are often grown in water that is shallower than necessary, and this practice has increased problems with a number of weeds, including watergrass, sprangletop, and smallflower umbrellaplant. Similarly, draining the field for any reason provides ideal conditions for germination and establishment of these weeds, many of which germinate or emerge poorly when water is 5 inches (13 cm) or deeper. Draining at or shortly after sowing may establish early uniform weed growth; however, draining multiple times will produce multiple weed flushes. This results in weed populations in many different growth stages, which are generally more difficult to control.

Careful land grading and seedbed preparation before planting help maintain uniform water depths in rice fields; and adequate canals, drains, and water control structures provide the means for efficiently moving irrigation water on and off the field. Laser planing allows for improved water management. Many rice fields are graded with a 0.02 to 0.05% slope, in contrast to row crop areas where a 0.1 to 0.2% slope is commonly used. Steeper slopes allow faster flood establishment, but flood depths can vary by several inches, impacting weed management.

Land leveling, grading, and efficient irrigation management are equally important to meet state-mandated water-holding regulations following herbicide applications. Nonuniform leveling can result in areas of deeper water, which can impact rice growth. Typically the field would be drained to remedy the rice growth impacted by deep water, but draining may have to be delayed due to state-mandated requirements for holding water.

Crop Rotation. Rotating rice with a dry fallow season or another crop is a useful technique for controlling a number of problem weeds, especially perennial weeds. Tilling to a depth of 8 to 12 inches (20–30 cm) during the fallow season can also aid weed management. Rotating to crops with different weed management strategies, such as tomato, safflower, or cereal crops, is one of the best ways to manage weeds that cannot be selectively controlled with herbicides and cultural practices in rice. However, heavy clay soil types limit rotation to other crops on more than half of California's rice acreage.

Rotation with rice can also help manage weeds such as johnsongrass, wild oats, nutsedge, fiddleneck, and field bindweed, which cause serious problems in upland crops.

Alternative Stand Establishment. Switching to an alternative stand establishment method in rice-only rotations may be used to manage weed populations, and may be especially useful for managing herbicide-resistant weeds. Weeds that build up in water-seeded rice can be greatly reduced by switching to dry-seeding for stand establishment, and weeds that build up in dry-seeded rice can be reduced by switching to water-seeding. The effectiveness of within-rice rotations to an alternative stand establishment is greatly increased by using the stale seedbed technique. This involves using preplant irrigation to encourage weed germination, then applying a nonselective herbicide with no residual activity to kill the weeds, after which fields are flooded and planted without additional tillage. However, it is important to hold the preplant floodwater long enough to allow the germination of obligate aquatic weeds such as smallflower umbrellaplant.

Herbicides

Herbicides are an integral part of a weed management program. Use them in combination with good crop management practices; mismanagement of water before and after herbicide application and failure to carry out preventive and cultural weed control methods reduce their effectiveness. Just as nonchemical methods alone are inadequate for satisfactory weed control, chemical methods rarely give complete weed control by themselves.

Each herbicide is effective on a different spectrum of weed species (see the *California Rice Production Workshop Workbook* and *UC IPM Pest Management Guidelines: Rice*). Extreme temperatures limit the effectiveness of all the major herbicides now recommended for use in rice. Base your choice of material on field-monitoring records from previous years, careful identification of weed seedlings in this year's crop, and monitoring of environmental conditions and crop developmental stages.

Most herbicides can be toxic to rice plants when used at excessive rates, under certain environmental conditions, or at certain stages of crop development. Postemergence applications of some materials injure rice in hot weather. Hot, dry winds can increase this injury when it is caused by foliar herbicides. Be sure to avoid combinations of materials or sequential applications of materials that will damage the rice plant; for example, applications of organophosphate insecticides (like malathion) within 14 days before or after an application of herbicides can result in serious crop damage; in the case of propanil, this is also true for carbamate insecticides like carbaryl.

Herbicide Types. Herbicides can be classified based on whether they are active on foliage or taken up from the soil by plant roots:

- *Soil-active herbicides* are adsorbed by the soil and are taken up by plant roots or emerging seedling shoots.
- *Foliar-active herbicides* must contact the foliage of target weeds to be effective. Some foliar-active herbicides are also active through the soil.

Certain herbicides are applied into the floodwater, often in granular form; some of them can be taken up by weed foliage in contact with the water and/or from the surface layer of soil by seedling roots or emerging shoots.

Herbicides can also be classified based on whether they move within the plant:

- *Translocated herbicides* move within the plant after being taken up by the roots or contacting the foliage.
- *Contact herbicides* move very little and kill only the part of the plant that is contacted by the spray.

Soil-active herbicides used in rice and some foliar-active herbicides are translocated. For foliar-active contact herbicides to be effective, uniform foliage cover by the spray is important, which requires flooded fields to be drained to expose small weed seedlings and allow herbicides to be applied with an adequate spray volume.

The types of herbicides used in rice are discussed in detail in the *California Rice Production Workshop Workbook* and *UC IPM Pest Management Guidelines: Rice.*

Mechanism of Action. Herbicides control plants by interfering with biochemical processes necessary for the plants' survival—for example, by blocking certain steps in photosynthesis. Most herbicides primarily affect one specific process, and this is called that herbicide's mechanism of action (MOA), sometimes also referred to as mode of action. More than one herbicide may have the same MOA. Tables in the *California Rice Production Workshop Workbook* and the *UC IPM Pest Management Guidelines: Rice* list the MOA categories for the herbicides used in rice. It is important to know the MOA when planning weed management strategies to prevent or manage herbicide resistance.

Herbicide Resistance. Herbicide resistance is the ability of certain biotypes within a weed species to survive and complete their life cycle after an herbicide treatment that kills all other biotypes of that species. Repeated use of the same herbicide or herbicides with the same MOA will select for herbicide-resistant biotypes (see Fig. 9, in the previous chapter).

Plants have different mechanisms that make them resistant to herbicides. Target-site resistance occurs when the biochemical pathway normally affected by the herbicide becomes resistant to the herbicide, usually through a point mutation in the gene coding for a protein to which the herbicide must bind to exert its toxic effect. Nontarget-site resistance is often the result of enhanced metabolic degradation (when a biotype becomes able to degrade the herbicide more rapidly than nonresistant biotypes). Or it can result from miscellaneous impairments in the ability of the herbicide to reach its site of action. Some weed biotypes may become resistant to more than one herbicide MOA.

Several factors favor the selection for herbicide resistance:

- excessive reliance on chemical control
- repeated use of herbicides with the same MOA and/or detoxification mechanism
- poor weed control, resulting in weedy fields, which increases the chance of encountering resistant biotypes
- weeds that produce large numbers of seeds with little dormancy and short longevity
- an herbicide with high efficacy on a specific weed species
- an herbicide with prolonged residual activity in the soil

Detecting Herbicide Resistance. When monitoring fields, watch for these signs of herbicide resistance:

- healthy-looking weeds alongside dead plants of the same species after an herbicide treatment
- no control of one susceptible species while other susceptible species in the same treatment area are controlled
- a gradual decline in control by a given herbicide
- discrete patches of a target weed species persistently surviving treatment with a given herbicide, with survival not attributed to misses during herbicide application
- resistance in a weed species that occurs in one field and is confirmed in a neighboring field

Special testing is needed to confirm herbicide resistance. Check with your farm advisor if you suspect herbicide resistance is developing.

Confirming Herbicide Resistance. The Weed Science Society of America (WSSA) and the International Survey of Herbicide Resistant Weeds (Survey; http://www.weedscience.org) have established five criteria that must be met for a weed biotype to be confirmed as herbicide resistant.

- Definition: The plant in question must meet the definition of herbicide resistance established by the WSSA and the Survey: "the inherited ability of a weed to survive and reproduce following exposure to a dose of herbicide normally lethal to the wild type."
- Testing: Resistance must be confirmed with a replicated, scientifically sound comparison of resistant and susceptible plants of the same species.
- Heritability: The resistance must be passed on to the progeny of the plant in question. Generally this is done by testing plants grown from seed produced by the presumed resistant biotype.
- Field impact: It must be demonstrated that the weed is a problem in fields where the herbicide is applied at the recommended rate.
- Identification: The plant in question must be a weed and must be identified to the species level.

Managing Herbicide Resistance. A number of cultural practices can be used to delay the selection for herbicide resistance or to manage weed populations that have been selected for resistance.

- Use weed-free certified seed and follow good sanitation practices to prevent introduction of weeds. This reduces the need for applying herbicides.
- Alternate rice stand establishment systems to help control weeds without use of herbicides; for example, dry-seeding may be used to manage herbicide-resistant aquatic weeds.

- Maintain an adequate water depth to suppress weeds.
- Use crop rotation whenever feasible.
- Do not rely only on chemical control.
- Rotate or mix herbicides with different MOAs.
- Control weeds that survive after an herbicide treatment, as this is critical. These plants are called escapes and have a high likelihood of being carriers of resistance genes; production of seed by these escapes must be prevented to avoid replenishing the soil with seed of herbicide-resistant weeds.

More information on managing herbicide resistance can be found in the references mentioned above and on the University of California Rice Research and Information Program website (www.plantsciences.ucdavis.edu/uccerice/).

Application Timing. Beyond choosing the material (or, in the case of sequential applications, combinations of materials), the most important factor in herbicide use is application timing. Herbicides must be applied only during certain stages of crop and weed development. At other times, these materials injure rice or give poor control of weeds. The application timing that maximizes weed control is specified on the herbicide label, and is commonly called the application window. It is always important to apply herbicides as early as possible relative to weed development. The longer weeds are allowed to grow, the more they compete with the rice crop and reduce yields.

Scheduling applications to occur a certain number of days or weeks after planting is not always reliable because the stages of crop and weed development can vary from year to year and field to field, depending on weather and other environmental factors. Thus, each field should be closely monitored for crop and weed development, especially during the first 5 weeks after the field is seeded. Proper times for applying major herbicides in relation to the growth stages of rice and target weeds can be found in the *California Rice Production Workshop Workbook* and *UC IPM Pest Management Guidelines: Rice*. In some years, the correct stage of crop growth and the correct stages of weed development may never occur simultaneously, and the herbicide gives less than optimal results—another reason not to rely on chemical control alone for managing weeds.

Use extreme caution when applying herbicides to control weeds later in the season. Losses can occur if herbicides are applied during rice panicle initiation and development. After the beginning of rice jointing, it is too late to control grasses with any herbicide.

Water Management. Controlling water movement and depth in the rice field is essential for obtaining optimal results from herbicide applications. When using foliar herbicides that are not translocated, weeds must be exposed as much as possible to be covered by the spray. This usually requires draining the field before application. Draining requirements should be specified on the herbicide label. If a field has been previously treated with an herbicide that requires a water-holding period before drainage, the water must be allowed to subside to the point where weeds are exposed before subsequent applications.

Water management strategies used for some grass herbicides depend largely on the requirements for holding water following their application. Check with your local agricultural commissioner to determine the water-holding requirements for your irrigation system. More information on water management strategies for herbicides can be found in the *California Rice Production Workshop Workbook* and the *UC IPM Pest Management Guidelines: Rice*.

Controlling Drift. Proper application procedures are essential for successfully using herbicides in rice and controlling drift. Damage to neighboring crops caused by herbicide drift is a potential problem with some herbicides. Drift can be controlled to a large extent by the size and position of nozzles as well as careful attention to formulation and environmental conditions.

The size of the herbicide droplets released by the application equipment is a major factor affecting drift. Small droplets drift farther than large droplets. However, a size compromise must be made because applications of very large droplets do not give good coverage and control. At present, the Microfoil nozzle, which can be used only on helicopters, controls drift the best. On airplanes, jet nozzles, with orifices of at least 1/16 inch (0.2 cm) and pointed backward parallel to the body of the plane, are best. You can get smaller droplets and better coverage by placing a whirlplate behind the jet nozzle. This is recommended for various insecticides but is not legal with the use of restricted herbicides in some areas. Do not use small whirlplates, as they increase drift to unacceptable levels.

Formulations also affect drop size. If properly used, thickening agents can increase average drop size, reducing but not eliminating drift; however, using them requires extra time and attention during equipment setup. If not properly used, they may actually increase drift. Also, adding thickening agents to sprays may decrease weed control because these materials increase drop size and reduce coverage.

Weather has a great influence on both the effectiveness of an herbicide application and its tendency to drift. Never apply an herbicide when the wind speed is greater than 15 miles per hour. With some materials and in some areas, 10 or 7 miles per hour is the legal limit. Avoid spraying materials when the wind is blowing toward sensitive crops or areas. A smoke column is used when aerially applying herbicides to detect changes in wind speed or direction. Also, whenever possible, avoid applying materials during air temperature inversions. Inversions are characterized by cool air at ground level, warm air overhead, stable and relatively low wind velocity, and unpredictable atmospheric conditions; even a light wind can shift an inverted air mass, transporting the pesticide away from the field.

Application procedures and conditions for restricted herbicides, including propanil, are strictly regulated by law.

Agricultural commissioners set any special requirements for applications of restricted-use materials within their jurisdiction. These requirements may include, but are not limited to, weather conditions, distance to sensitive crops, and type of application equipment. Check with your county agricultural commissioner for local requirements.

Monitoring and Identifying Weeds

Accurate identification of weed species and careful field monitoring are necessary for choosing the best weed management program, especially when using an herbicide. Most of the herbicides used in rice are effective on only one or a few species, so correct identification is critical for choosing the right material. Also it is important to identify herbicide-resistant weed populations as soon as possible.

Two kinds of monitoring operations are required: monitoring weed species present and their densities, and monitoring growth stages of rice and certain problem weeds. You should survey each field, preferably check by check, for weed species and densities at least twice during the growing season—between flooding and 45 days after seeding and between panicle initiation and heading. Use records from these surveys to plan your weed control and crop rotation programs. An example of a survey form can be found in the *UC IPM Pest Management Guidelines: Rice*. Include a map of each field, showing areas where perennials and other troublesome weeds are located. Monitor the development stages of rice and problem weeds during stand establishment. Monitoring growth stages is essential to control weeds effectively with herbicides and avoid damage to the rice plant. Begin monitoring as soon as fields are flooded or dry-seeded.

The major weed species found in California are listed in Table 2. The most common or troublesome are pictured in this book. They are arranged according to common location or habitat. These habitats are depicted in Figure 10. Field weeds are further divided according to growth habit—emersed, floating, or submersed. Emersed weeds may also be a problem in shallow ponds and drainage canals; nearly all stand erect and are rooted in moist or flooded soil. Floating weeds rest on the water surface and rise or fall with water level fluctuations; their roots may hang freely in the water or be established in the soil. Submersed weeds include the pondweeds and naiads that grow largely underwater, occasionally with some floating leaves or flowering stems a few inches above the surface. Weedy algae begin growth on the soil. Eventually they float to the water surface, forming large mats and inhibiting rice seedling emergence. Figure 11 illustrates some of the terminology used in describing weeds.

Aquatic and Riparian Weeds of the West and *Weeds of California and Other Western States*, listed in the suggested reading, illustrate additional species and give more complete botanical descriptions. Photos and descriptions of additional species can also be found in the "Weed Photo Gallery" associated with the *UC IPM Pest Management Guidelines* on the UC IPM website (www.ipm.ucdavis.edu). Your farm advisor can assist you in identifying weeds or refer you to someone else who can.

Watergrass and Barnyardgrass

***Echinochloa* spp.**

Watergrasses and barnyardgrass are the most serious weeds of rice in California. Although close relatives, these weeds are quite distinct and in recent years have been classified into three separate species: early watergrass *(Echinochloa oryzoides)*, late

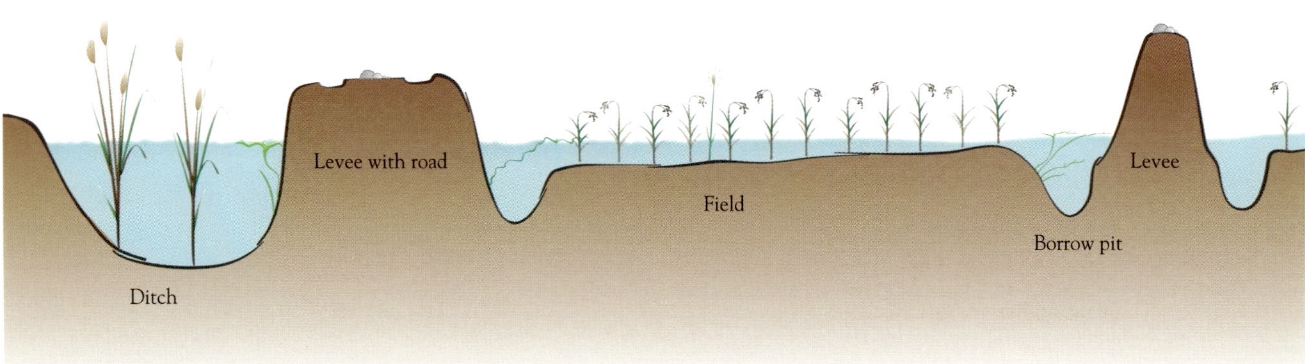

Figure 10. Weed habitats in the rice field (not to scale). Some permanent basins may not have borrow pits.

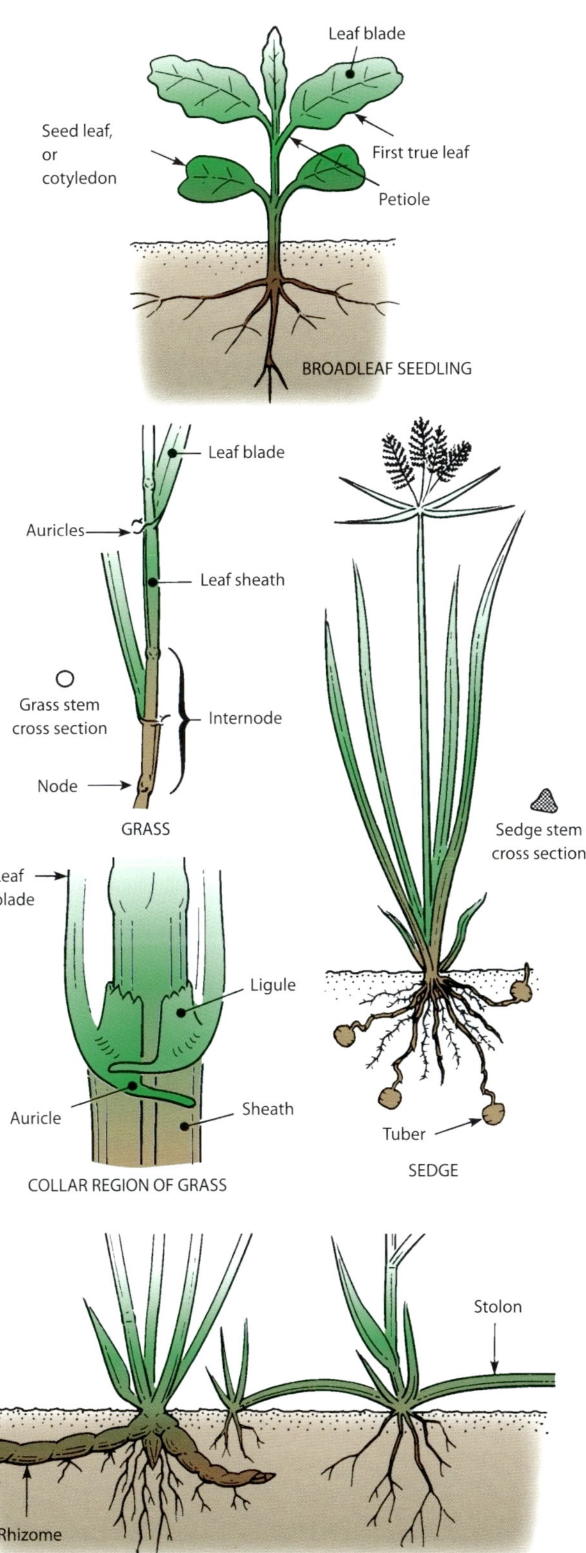

Figure 11. Vegetative parts commonly used to identify weeds. Definitions of terms can be found in the glossary.

Table 2. Common Weeds in California Rice Fields, Classified by Habitat.

LEVEE
cattail
barnyardgrass
johnsongrass
knotgrass
smartweed
sprangletop
BORROW PIT
cattail
ducksalad
knotgrass
smartweed and most field weeds
FIELD
Emersed
burhead
California arrowhead
cattail
ducksalad
early watergrass
Gregg arrowhead
late watergrass
monochoria
redstem
river bulrush
ricefield bulrush
smallflower umbrellaplant
spikerushes
waterplantain
Floating
Eisen waterhyssop
roundleaf waterhyssop
Submersed
American pondweed
blue-green, unicellular algae
Chara
green filamentous algae
horned pondweed
southern naiad
DRAINAGE DITCH
duckweed
cattail
hardstem bulrush
water primrose

watergrass (*E. phyllopogon* = *E. oryzicola*), and barnyardgrass (*E. crus-galli*). The early and late reference to the watergrasses is in relation to their flowering times; they germinate at about the same time. Early watergrass flowers about 40 days after flooding, well ahead of rice; the heads are drooping (pendant), and the seed (spikelets) are heavily awned. Late watergrass flowers about 90 days after flooding, at or near the time of early rice varieties; the heads are erect and the seed are awnless. Its growth habit closely resembles that of rice. Both early and late watergrass have larger seed than typical barnyardgrass and are more successful in emerging when fields are continuously flooded. Barnyardgrass is especially prevalent in intermittently irrigated (dry-seeded) or shallow-water rice cultures. Barnyardgrass is relatively small-seeded, quite variable in appearance, and is a weed in a wide range of irrigated crops, including rice. Unlike barnyardgrass, early and late watergrasses are only weeds of rice and can emerge and grow under flooded conditions.

From 1914 to about 1930, California rice was dry-seeded, and barnyardgrass infestations became so heavy that many fields were abandoned. The California system of water-seeding was established principally as a means to control barnyardgrass, and continuous flooding still provides good control. In fields flooded continuously with a water depth of 5 inches (13 cm) or more, generally only those plants on levees or that were established before flooding survive.

The larger-seeded watergrasses are better adapted to continuous flooding and are now more commonly found in rice fields. Early watergrass is the most widespread, occurring throughout the rice-growing areas in California, whereas late watergrass is more localized in the northern Sacramento Valley. Water depths less than 5 inches (13 cm) cannot be relied upon to control watergrasses, but 7 to 8 inches (17–20 cm) of water can provide about 50% control.

Watergrass and barnyardgrass seedlings can be controlled by preflood cultivation if they are dislodged before secondary roots have begun to grow. Maintain a continuous water depth of at least 4 inches (10 cm) during rice stand establishment, in combination with the use of appropriate herbicides, to limit infestations of these weeds. Proper timing is critical for success (see the *California Rice Production Workshop Workbook* and *UC IPM Pest Management Guidelines: Rice*). Very severe infestations may require rotation to another crop where possible or use of an alternative stand establishment method. Resistance to several rice herbicides has become widespread in populations of *Echinochloa* species.

A. A mature watergrass plant grows upright, although it may spread at the base, and is stout, varying from 6 inches to 6 feet (up to 1.75 m) in height. Under favorable growing conditions, it roots and tillers at the lower nodes to form large clumps.

B. The best way to distinguish watergrass and barnyardgrass from rice and other grasses is to pull back the leaf and look at the collar region. The *Echinochloa* species are one of the only weedy grasses that do not have a ligule at the junction of the collar and blade. This characteristic is useful for identifying watergrass and barnyardgrass seedlings and mature grasses. The plant at right is rice.

C. Flowering heads, or panicles, of barnyardgrass are highly variable and may be tinged purple, long awned, short awned, or lacking awns. Flowering heads may be up to 10 inches (25 cm) in length. They have smaller seeds than either early or late watergrass and thus do not droop like the other two varieties.

D. Drooping flowering heads with long awns are more typical of early watergrass. This variety is widespread and is usually found heading midseason.

E. Short awns to awnless, large seeds, and erect to slightly drooping heads are typical of late watergrass. It is found in the middle of fields, flowering late in the season in localized areas.

Red Rice

Oryza sativa

Red rice is currently considered a noxious pest. In the past few years, red rice has been found in the Sacramento Valley. It had been prevalent in the Sacramento Valley before the introduction of certified seed and water-seeding in the 1930s but had largely disappeared under the culture of continuous flooding. It most likely evolved through hybridization between cultivated rice and the wild rice *Oryza rufipogon* in California.

Red rice is a form of weedy rice that is characterized by the presence of a red pericarp, which confers an undesirable dark reddish color to the grains. The plant is difficult to distinguish from rice cultivars that have red grains, except that the seed of red rice shatters readily from the panicle before maturity.

The introduction of other types of red rice from the southern United States, where it is a major problem in rice, could eventually occur. Eradication efforts should be implemented immediately if red rice is found. If you find plants you suspect may be red rice, contact your farm advisor or county agricultural commissioner's office.

F. The panicle of red rice (left) is compact and erect, and the florets have long awns. The panicle of cultivated rice (right) is open and drooping, and the florets have no awns or very short awns.

G. Seeds of red rice (right) are reddish brown, in contrast to the light-colored seeds of most rice cultivars (left).

Bearded Sprangletop

***Leptochloa fascicularis* (=*Leptochloa fusca* var. *fasicularis*)**

Bearded sprangletop is an annual grass that frequently infests rice fields as well as irrigated crops, orchards, vineyards, and irrigation and drainage canals. It is particularly troublesome where rice stands are thin or where water has been drained,

A. Mature watergrass plant.

B. Distinguish watergrass and barnyardgrass from rice and other grasses by pulling back the leaf and looking at the collar region. The plant at right is rice.

C. Flowering head, or panicle, of barnyardgrass.

D. Early watergrass.

E. Late watergrass.

F. Red rice (left) and cultivated rice (right). *Photo:* Larry L. Strand.

G. Seeds of red rice (right) in contrast to light-colored seeds of most rice cultivars (left). *Photo:* Larry L. Strand.

allowing it to germinate. Bearded sprangletop seedlings do not survive well underwater. A well-cultivated and -prepared seedbed and continuous water levels of 4 inches or more (over 10 cm) can limit sprangletop infestations. Severe infestations may require a rotation with another crop.

H. Bearded sprangletop plants thrive in rice fields, sometimes attaining a height of 4 feet (1.2 m) and often growing in large clumps. The color of the foliage is generally lighter in rice fields than in plants growing in orchards and vineyards; nitrogen levels can also cause considerable color variation.

I. You can distinguish seedlings and mature bearded sprangletop plants from any other grasses and rice by examining the collar region. The sprangletop ligule is thin and membranous and tears easily. The sheath and leaf blades of mature plants are somewhat rough, and the lower sheaths may be straw colored to reddish. Seedling leaves usually have a white midrib.

J. Young flowering heads of bearded sprangletop are dark green and become lighter green as they mature. After shedding seed, the heads turn straw colored. Heads are branched, erect, and 6 to 16 inches (15.2–40.6 cm) long. Each branch is about 1 to 6 inches (2.5–15.2 cm) long with several spikelets. Spikelets are short stemmed, containing 7 to 11 flowers. Bearded sprangletop can be distinguished from the other sprangletops that grow in California because it is the only one with awns on the tip of each flower. These awns are quite short.

Rice Cutgrass

Leersia oryzoides

Rice cutgrass is a perennial grass native to California. It grows in wet soil conditions, so rice culture provides a good habitat. Rice cutgrass has rough, finely-toothed leaf edges, which can injure people by cutting flesh. This grass is a problem weed in rice-growing areas of the southeastern United States and has been found in some areas of the Sacramento Valley. Rice cutgrass spreads primarily by rhizomes and can be controlled by tillage that chops up the rhizomes and allows them to dry out. Most herbicides available for rice do not control this weed; repeated applications of glyphosate in late summer will control infestations.

K. Rice cutgrass grows up to 5 feet (1.5 m) tall. Panicles are up to 8 inches (20 cm) long with spikelets that are ⅙ to ¼ inch (4–6 mm) long. Each spikelet produces a single, reddish brown grain. Young panicles are often only partially exerted from the uppermost leaf.

L. The ligule of rice cutgrass is short and papery. The sharp serrations on the leaf margins can cut through human skin, giving the plant its name.

Smartweeds

***Polygonum* spp.**

Smartweeds are widely distributed in marshy areas, along shorelines of rivers, streams, lakes, ponds, and irrigation and drainage ditches. Although it does not require flooded conditions to complete its life cycle, smartweed is well adapted to the aquatic environment; roots and shoots can develop at stem joints lying in water. In rice, it grows primarily on levees and in open water at the field edges, although it is an increasing problem in fields with shallow water, in dry-seeded rice, and in fields that are drained. It is associated with cold water, often growing around cold water inlets. Two species, pale smartweed, *Polygonum lapathifolium*, and ladysthumb, *P. persicaria*, are commonly found in rice. They reduce yields, interfere with harvesting and drying, and contaminate rough and milled rice. Smartweeds are partially controlled by mechanical cultivation of levees in the fall and by herbicides applied to control other broadleaf weeds.

Pale smartweed plants may be erect and 4 to 5 feet (1.2–1.5 m) tall. Ladysthumb may be erect or creeping with weak ascending stems. Stems are red and swollen at the nodes. The leaves are long and narrow, tapering at the tip, with whitish undersides. Pale smartweed flowers range in color from greenish to pale pink. They are borne on stalks in long, dense, drooping clusters. Leaves of ladysthumb have purple blotches in their centers, and its flowers are pink and erect.

M. Smartweeds often grow in open water or along field edges.

N. The smartweed seedling has dull green, oval seed leaves, which are about 1½ to 2 times as long as they are broad. The early true leaves are larger and more pointed. The curled leaves emerge in a distinguishing manner: they tear the membranous tubular sheath at the base of the previous leaf.

Cattails

***Typha* spp.**

Cattails frequently invade irrigation and drainage ditches and may grow in rice where drainage is inadequate. They compete with rice, impede the flow of water in irrigation systems, and slow harvesting. Cattails are perennials that reproduce both by seed and by rhizomes. Seed germinate readily in mud and

H. Bearded sprangletop plants.

I. Collar region of bearded sprangletop.

J. Young flowering heads of bearded sprangletop.

K. Rice cutgrass.

L. Ligule of rice cutgrass.

M. Smartweeds.

occasionally in shallow, clear water. Because of their extensive rhizome systems, established cattail colonies are difficult to control without deep tillage followed by a fallow period or a rotation with another crop. Rhizomes must be desiccated to be killed.

Cattails emerge from seed or rhizomes and grow upright, developing thick, light green leaves. The lower part of the first leaf is whitish. A cattail with two or three leaves can be easily distinguished from grasses by making a cross section of its round stem, which is filled with white, spongelike pith. Grass stems do not contain spongy pith and are hollow at their nodes. The cylindrical stem distinguishes cattails from most sedges, which have stems that are triangular in cross section; tule and spikerush are sedges that have cylindrical stems.

O. Cattail infestations start in drainage ditches or other areas of shallow water and, if not carefully managed, may spread by rhizomes into the rice field. Cattails can also spread into the field when rhizomes are transported during plowing.

P. Cattail plants are 4 to 8 feet (1.2–2.4 m) tall and usually grow in groups. The flowering stem grows from the base of the plant and may become as long as the leaves. Flowers are held at the tip of the stem in a dense cigar-shaped cluster about 4 to 7 inches (10–15 cm) long. Cattails are sometimes confused with hardstem bulrush, *Schoenoplectus acutus* (=*Scirpus acutans*; locally called tule), a very tall (up to 15 ft, or 450 cm) sedge occasionally found growing in rice fields or drainage ditches. However, hardstem bulrush plants consist primarily of cylindrical stems, with only a short (3 in, or 7.5 cm) leaflike blade and sheath at the base of the stem. Cattail leaves are somewhat flat and tend to twist. Mature bulrush stems end in a flower that is sedgelike and very different from the cattail flower.

Arrowheads

***Sagittaria* spp.**

Two species of arrowheads cause problems in California rice. California arrowhead, also called giant arrowhead, *Sagittaria montevidensis* ssp. *calycina*, is a native annual aquatic plant in the waterplantain family and is widely distributed in rice fields and other marshy areas. Gregg arrowhead, *S. longiloba*, (also known as longbarb arrowhead or long lobed arrowhead) is a native perennial, reproducing from seed and rhizomes, and typically distributed in patches within the field. Seedlings of both species grow completely submerged in water, but mature leaves and flower-bearing stems grow above the water surface. Arrowheads are primarily a problem in areas where the rice stand is thin or in borrow pits.

Q. Arrowhead seedlings often grow completely underwater. The succulent seed leaves are narrow and light green with nearly parallel sides that taper to a blunt point. The first true leaves are similar in shape although slightly wider at their base. Rectangular-shaped markings on the leaves distinguish arrowheads from similarly shaped seedlings of ducksalad and bulrush. At this stage, it is impossible to differentiate the two arrowhead species. The characteristic arrowhead leaf shape does not appear until the third or fourth true leaf.

R. The mature leaves of arrowheads are distinctive and easy to recognize in the rice field. They are arrowhead shaped with widely spreading pointed lobes on either side of the leaf stalk; the length of the lobes is quite variable. Each leaf stalk rises from the base of the plant and bears a single leaf. The mature leaves of Gregg arrowhead (background) are much narrower than California arrowhead leaves (foreground) and have basal lobes two to three times longer than the terminal lobe. The flowers in this picture are Gregg arrowhead. The flower stalk of California arrowhead does not grow above the leaves.

S. These are staminate (male) flowers of California arrowhead. Flowers are borne in groups on leafless stems. Flowers lower on the stem have both male and female parts, but those higher up, such as those shown here, have only male parts (stamens). The round green structure at right is a flower bud; stalks bearing fruit droop, and the surface of the fruit is granulated. The insects on the bud stalk are aphids.

T. Rhizomes of Gregg arrowhead often grow out from the base of the plant to form globular perennial bulbs (corms) from which new plants arise. Rhizomes are spread by cultivation and waterfowl, and cannot be killed by herbicides currently registered for use in rice.

Common Waterplantain

Alisma plantago-aquatica

Common waterplantain is a native perennial aquatic plant that infests rice fields, especially in areas where the stand is sparse and where cold water enters the field. Waterplantain can regrow from its bulbous base, creating problems where fields have been inadequately tilled, but in most fields, plants originate primarily from seed.

N. Smartweed seedling.

O. Cattail infestation.

P. Cattail plants and flowers.

Q. Arrowhead seedling.

R. Mature leaves of arrowheads.

S. Staminate (male) flowers of California arrowhead.

T. Rhizomes of Gregg arrowhead.

U. Unlike arrowhead seedlings, waterplantain seed leaves usually float on the water surface. They are elliptical, about 2½ to 3 times as long as broad, and have a long reddish-tinged stalk. The first true leaves take on the typical waterplantain shape but are much smaller than mature leaves.

V. The mature plant grows from a bulbous base to a height of 3 to 4 feet (90–120 cm). A leafless stem, shaped like a pyramid and bearing flowers, extends well above the leaf blades. Occasionally a plant has more than one flowering stem. Flowers have three petals and are white, green, or pink and held in whorls of 3 to 10. Leaves are prominently veined and elliptical with no basal lobes and are about 2 to 8 inches (5–20 cm) long and about half as broad. Each leaf stalk bears a single leaf. Leaves are difficult to distinguish from those of burhead.

Upright Burhead

Echinodorus berteroi

Upright burhead is an emerged native annual to short-lived perennial aquatic plant in the waterplantain family that grows in shallowly flooded areas such as irrigation drainage canals and in rice fields where the stand is thin. Like waterplantain, upright burhead mostly grows from seed.

W. Prior to flowering, upright burhead can be confused with waterplantain. Like waterplantain, its leaves are elliptical, and each leaf is borne singly on a leaf petiole that arises from a bulbous base rooted in the soil. However, upright burhead leaves have wavier margins than waterplantain leaves, and the leaf bottoms are more heart shaped. Upright burhead leaves often have significant symptoms of insect damage. The leaves in this picture are typical. Leaves are up to 10 inches (25 cm) long and about half as wide. Flowering upright burhead plants are easy to recognize because the flowers and burrs that are borne on the tall flowering stems are quite distinctive.

X. Upright burhead flowers are held in whorls and have three widely spaced white petals. The fruit is a distinctive burr that turns brown as it matures.

Ducksalad, Waterstargrass

***Heteranthera* spp.**

Ducksalad, *Heteranthera limosa*, and waterstargrass, *H. dubia*, are emersed or submersed aquatic plants in the pickerelweed family (Pontederiaceae). Their seeds germinate only in saturated soil. They are a problem in rice primarily in open water early in the season. Waterstargrass is a native perennial found in the Sacramento Valley, Modoc Plateau, and along the North Coast. Ducksalad is a non-native annual with a more limited distribution, found mostly in the Sacramento Valley. Two other non-native *Heteranthera* species have been recently found in California. *Heteranthera* species can be managed with currently available broadleaf herbicides.

Y. Ducksalad, *Heteranthera limosa*, plants have showy flowers with six narrow, white petals arranged in the shape of a star. Each flower is borne singly on a flowering stalk. The leaves are oval, much smaller (½–5 in, or 1–13 cm) than waterplantain leaves, and are waxy green. Leaves may be submerged, floating, or held above the water surface. Stems of this species are rarely elongated.

Z. Roundleaf mudplantain, *Heteranthera rotundifolia*, is a newly discovered non-native species with blue or white flowers. Stems are frequently elongated and may root at the nodes. This species is more aggressive than ducksalad. It has a tendency to form dense mats that are difficult to control by drying down fields.

AA. Bouquet mudplantain, *Heteranthera multiflora*, has been recently identified in only a few locations in the Sacramento Valley. This species has a similar growth habit to roundleaf mudplantain, *H. rotundifolia*, but is distinguished by having multiple mauve to white flowers on each flowering stalk (A). The other *Heteranthera* species have a single flower per flowering stalk. Leaves (B) are rounder than those of ducksalad, *H. limosa*.

U. Waterplantain seed leaves.

X. Upright burhead flowers.

V. Mature waterplantain.

Y. Ducksalad plants.

W. Upright burhead.

Z. Roundleaf mudplantain. *Photo:* James W. Eckert.

AA. Bouquet mudplantain. This species has a similar growth habit to roundleaf mudplantain but is distinguished by having multiple mauve to white flowers on each flowering stalk (A). The other *Heteranthera* species have a single flower per flowering stalk. Leaves (B) are rounder than those of ducksalad. *Photos:* James W. Eckert.

Monochoria

Monochoria vaginalis

Monochoria is an aquatic plant in the waterplantain family that infests some rice fields in the Sacramento Valley. It grows as a perennial under constantly flooded conditions. *Heteranthera rotundifolia* is sometimes confused with monochoria, but the leaves and flowers have distinct differences. Tillage of dry fields will destroy monochoria rhizomes, but this plant also reproduces from seed. Monochoria tends to germinate and establish later in the season than *Heteranthera* and remains present into rice harvest. Seed dispersal is late in the season after drydown, which may explain why it is not spreading as rapidly as ducksalad, which disperses seed midseason while rice is still flooded. Although this plant is not widely distributed in California (only found in the Sacramento Valley), it is on the Federal Noxious Weed List.

BB. The leaves of young monochoria plants (A) are narrow, elongated, and lance shaped. These narrow, pointed leaves are distinctly different than leaves of ducksalad. Leaves of older monochoria plants (B) are much broader than those of young plants. They are rounded or heart shaped at the base, with pointed tips.

CC. The blue to violet flowers of monochoria are borne close to the stem, with an elongated bract at their base. The presence of flowers in a cluster and the prominent bract distinguish monochoria from blue-flowered *Heteranthera*.

Redstem

Ammannia spp.

Two species of *Ammannia* occur in California, *A. coccinea*, called redstem, and *A. robusta*, termed purple ammannia. However, both of these native species are extremely difficult to distinguish, and most growers use the name redstem for both species. Both are grouped here as redstem. Redberry is another common name for these weeds. The two species differ only in the presence or absence of small stalks attaching the flowers to leaf axils; their biology and control are similar. Redstem is an aquatic, emerged annual that reproduces by seed. The weed grows in most wet areas and does especially well in rice during warm years and in fields where the stand is thin. Like many other broadleaf weeds, the best way to manage redstem is to maintain a dense stand. Herbicides are effective against redstem, but redstem plants often germinate after the early-season herbicide treatments have lost their residual effectiveness, causing the weed to appear in places where the rice stand is poor, about the time rice plants are heading. Redstem lowers rough rice quality and the price paid to the grower because its seedpods give a false high-moisture reading for the harvested crop.

DD. The seed leaves on redstem seedlings are triangular, dark green, often reddish tinged, and 1½ to 2 times as long as broad. The first pair of true leaves are opposite and oval or triangular.

EE. Mature redstem plants may be from 5 to 30 inches (12.5–75 cm) tall. The stems branch extensively, especially where rice is sparse. The narrow, pointed leaves clasp the stem in opposite pairs, with the next set of leaves in a perpendicular arrangement. As the plant matures, the stems and leaves turn bright red.

FF. Small flowers with four petals grow in groups of three or four at the base of the leaves. The flowers are usually bright rose red. Flowers that lack stalks are *Ammannia coccinea* (inset), the most common *Ammannia* species in California.

Bulrushes

Schoenoplectus spp.

Two species of *Schoenoplectus* are found in California rice—*S. fluviatilis*, river bulrush, and *S. mucronatus*, ricefield bulrush. River bulrush is a native perennial sedge that occurs in drainage and irrigation canals and can be troublesome in rice fields not rotated with other crops. Infestations are localized but can be problematic where they occur. River bulrush grows vegetatively from tubers and appears very early in the season, usually in clumps or patches. This distinguishes it from the more widespread ricefield bulrush, a more damaging non-native weed that generally functions as an annual and does not usually flower until midseason, about 60 to 70 days after flooding. On occasion, it also grows as a weak biennial or perennial.

Ricefield bulrush plants are easily controlled, but river bulrush plants are not. Freezing or tilling can aid in the control of river bulrush.

GG. Seedlings of ricefield bulrush can be distinguished from those of smallflower umbrellaplant and most other sedges by the angle at which the leaves grow out. When viewed from above, leaves on ricefield bulrush seedlings appear to grow out only in two directions at 180° angles, or opposite each other.

HH. River bulrush has extensive underground stems (rhizomes) that form dark brown tubers, from which new plants may grow. Tubers are very hard and difficult to cut with a pocket knife. Deep-plowing fields to a depth of 8 to 12 inches (20–30 cm) followed by a rotation to a nonirrigated crop or to fallow is the best way to control river bulrush once tubers are established.

BB. Leaves of young monochoria plants (A). Leaves of older monochoria plants (B). *Photo:* James W. Eckert.

CC. Flowers of monochoria. *Photo:* James W. Eckert.

DD. Seed leaves on redstem seedlings.

EE. Mature redstem plants.

FF. Bright rose red flowers of redstem.

GG. Seedlings of ricefield bulrush.

HH. River bulrush rhizomes.

II. With a typical height of 3 to 5 feet (90–150 cm), river bulrush is taller than most other sedges but has the three-sided stem characteristic of many members of the sedge family. Several slender tapering leaves alternate on the stem, and the stems terminate in a flowering head with three to five secondary leaves just below. The flowering head is loose with several flower clusters. The clusters are egg shaped and pointed at the tip.

JJ. Ricefield bulrush can be distinguished from river bulrush because it has only one leaf just below its flowering head, while river bulrush has three to five. The leaf tips point away from the flower at an angle as the flower blooms. Ricefield bulrush plants are generally shorter than river bulrush, growing to a height of about 2 to 3 feet (60–90 cm).

Smallflower Umbrellaplant

Cyperus difformis

Smallflower umbrellaplant is a non-native sedge that is distributed throughout the rice-growing areas. The plant produces thousands of small seed that become uniformly distributed over the field, resulting in very high seedling populations. Smallflower umbrellaplant can be managed with proper herbicide application.

KK. Leaves of smallflower umbrellaplant seedlings grow out in three directions at approximately 120° angles.

LL. Smallflower umbrellaplant flower clusters are globular and may or may not be on stalks. The individual flowers are greenish white with a brown marking on each side. Stems are triangular in cross section.

MM. As the flowers mature into fruiting heads, they turn brown. Two or three secondary leaves grow out of the flowering stem just below the flower clusters.

Spikerushes

***Eleocharis* spp.**

Spikerushes are natives that grow on levees, in shallow ditches, on poorly drained soil, and in rice fields. They compete with rice early in the growing season. Like bulrushes and smallflower umbrellaplant, they belong to the sedge family but have stems that are round rather than triangular in cross section.

NN. The stems of mature spikerush plants terminate in a dense flowering head. Mature plants have no leaves. This is blunt spikerush, *Eleocharis obtusa*; plants are short, rarely growing more than 1½ feet (0.5m) tall.

Waterhyssops

***Bacopa* spp.**

Waterhyssops form a floating mass of leaves and flowers on the water surface. They invade rice fields early in the season, often interfering with stand establishment. Four species occur in California rice fields; all are annuals. Eisen waterhyssop (*Bacopa eisenii*) is native and more difficult to control. It is found primarily in the San Joaquin Valley. Monnier waterhyssop (*B. monnieri*), creeping waterhyssop (*B. repens*), and disk waterhyssop (*B. rotundifolia*) are less common non-native species.

Waterhyssop plants have many branched and submerged stems that may develop roots at the nodes. Leaves occur in pairs opposite each other on the stems; they are not stalked and are about ½ to 1 inch (1.0–2.5 cm) long, nearly circular, or slightly wedge shaped. Flowers have five white petals and a yellow center.

OO. The flowers of Eisen waterhyssop, *Bacopa eisenii*, are almost as big as the leaves and are quite showy.

PP. Flowers of disk waterhyssop, *Bacopa rotundifolia*, are smaller. When in bloom, the flower is held above the water, but after the fruit matures, the flowering stalk sinks beneath the water.

QQ. The seed leaves of waterhyssops are light green, lance shaped, and 1½ to 2 times as long as they are broad and grow on the basin bottom. The first pair of true leaves are submerged and appear on stems that are short or elongated, depending on the water depth. They are elliptical to round and much broader than the seed leaves. The second pair of true leaves is usually floating.

II. River bulrush.

JJ. Ricefield bulrush.

KK. Leaves of a smallflower umbrellaplant seedling.

LL. Smallflower umbrellaplant flower clusters.

MM. Flowering stems of smallflower umbrellaplant.

NN. Blunt spikerush.

OO. Flowers of Eisen waterhyssop.

PP. Flowers of disk waterhyssop.

QQ. Seed leaves of waterhyssop.

Pondweeds

Potamogeton spp.

Pondweeds are aquatic perennials that usually grow in ditches and reservoirs, but occasionally they are found in rice. Pondweeds reproduce by seeds, rhizomes, and tubers. Tilling to a depth of 8 to 12 inches (20.3–30.5 cm) followed by a rotation to a nonirrigated crop provides effective control.

RR. Pondweed plants have both floating and submerged leaves that may be quite different in appearance. Submersed leaves are alternate, but the upper leaves may be opposite. Stems are simple or slightly branched. Leaf structure varies from species to species, from grasslike to broadleaf with many distinct veins. The species shown here is American pondweed, *Potamogeton nodosus*. Pondweed flowers are in elongate heads or spikes held above the water surface.

SS. American pondweed produces winter buds on the tips of rhizomes, usually beneath the soil surface. Control is most effective before winter buds form. The winter bud shown here has been pulled from the mud and is just beginning to sprout.

Naiads

Najas spp.

Naiads are rooted, submersed aquatic annuals that reproduce by seed and vegetatively through propagation of stems. They compete with rice in the early growing season, especially in fields with deep water, and they may reduce tillering. Naiads may also interfere with mosquito control by acting as a barrier to mosquito fish. Surface-applied nitrogen and phosphorus fertilizers that are not incorporated may enhance naiad growth.

TT. Infested rice fields may have masses of naiad stems submerged and floating in areas where the stand is thin. Stems are 10 to 25 inches (25–65 cm) long and may branch at the nodes.

UU. Leaves encircle the stem in whorls of three. They are very narrow (0.01–0.1 in, or 0.25–2.5 mm wide) and taper at the tips. Tiny axillary flowers emerge from the sheath at the base of the leaf.

Algae

Algae are nonflowering, plantlike organisms that reproduce by cell division or through the production of spores. The types that commonly occur in rice fields include

- phytoplankton, which grow suspended in water as single cells or in microscopic colonies
- filamentous and colonial algae, which grow in long strands or filaments large enough to be seen with the unaided eye
- *Chara*, which resembles a higher plant

The more troublesome are the filamentous and colonial algae. Although phytoplankton in large numbers may cloud the water, they rarely cause economic losses. *Chara* algae cause problems only in highly alkaline fields.

Filamentous and colonial algae can clog waterways and impede the growth of young rice plants. Masses of filamentous and colonial algae may uproot seedlings. Algae grow well at all water depths that are common in rice fields, but are especially troublesome where poor seedbed preparation has created low areas with poor surface drainage and in fields with inadequate decomposition of organic matter. When adding phosphorus to fields that will be water-seeded, incorporate the fertilizer at least 2 to 4 inches (5.1–10.2 cm) deep before flooding. Phosphate fertilizer left on the soil surface or applied shortly after flooding may stimulate the growth of algae, and especially the cyanobacterium *Nostoc spongiaeforme*, often referred to in the field as black algae. In cases where *N. spongiaeforme* is expected to be a problem, delay phosphorus applications until after the rice stand has established to prevent stimulation of growth.

Various copper compounds are available to control filamentous algae. To be effective, these materials must be applied before algae mats float to the water surface. The presence of rice straw decreases the effectiveness of copper compounds because copper binds to the straw. Check for bubbling on the water surface, which is an indicator of a serious algal infestation.

VV. Filamentous algae are most abundant in newly flooded rice fields early in the season. The algal units unite to form a solid mass of scum suspended from the water surface by gas bubbles and extending to the bottom of the flooded basin.

WW. This photo shows an algal mass that has eliminated much of a rice stand by pulling rice plants down into the water, creating a scum barrier that young rice tillers cannot penetrate, and blocking the transmission of sunlight to submerged leaves.

XX. Anchored into the mud, *Chara* plants are branched with evenly spaced whorls of leaflike structures at the nodes. They may be mistaken for more advanced plants like naiads. When crushed, *Chara* plants emit a musky garlic or skunklike odor. Plants are usually gray-green but may turn brown or chalky due to calcareous deposits or accumulated dead plants. They are usually found associated with alkaline soil of a pH 8 or higher. *Chara* persists through the season, holding water and preventing soil from drying at harvest. No satisfactory controls are currently available, although temporarily draining the field may limit an infestation.

RR. American pondweed.

SS. American pondweed winter buds on the tips of rhizomes.

TT. Masses of naiad stems.

UU. Naiad leaves encircling the stem in whorls of three.

VV. Filamentous algae.

WW. An algal mass that has eliminated much of a rice stand.

XX. *Chara* plant.

Photo: Mike Poe

Invertebrates

Rice water weevil generally is the most economically damaging invertebrate pest in California rice fields (Table 3). Rice seed midges, tadpole shrimp, and crayfish can also cause serious damage. In California the most serious invertebrate problems, and thus the period of most management activities, occur in the first month to 6 weeks after planting. Tadpole shrimp, crayfish, and rice seed midges feed on germinating rice seed and small seedlings. This feeding can kill rice plants or lead to a general weakening of the plants. The rice leafminer attacks floating leaves of young plants, but once leaves start growing upright above the water, leafminers do not damage them, and the pest ceases to be of economic concern. Although the rice water weevil can feed on rice plants throughout the season, economic damage is caused by larvae feeding on young plants, and treatments must be made during the early weeks of stand establishment in order to avoid economic losses.

Older plants are not typically damaged by invertebrate pests. Once the stand is well established, armyworms and leafhoppers are the only invertebrates that may require management, and neither is found at damaging levels in most fields every year. Thus, the period of most intense activity in managing invertebrate pests is the period between sowing seed and stand establishment.

Although they do not damage rice, mosquitoes are of concern to growers planning an IPM program. Rice fields provide an ideal breeding habitat for mosquitoes, and growers have a legal responsibility to limit mosquito populations as much as possible.

Table 3. Relative Importance of Invertebrate Pests in California Rice.

MAJOR PEST Occurs in most fields every year; monitoring and management action needed every year	MINOR PEST Occurs in most fields most years; monitoring needed every year, management action some years	OCCASIONAL PEST Occurs in some fields some years; some monitoring needed every year, management action only in occasional years
rice water weevil	rice seed midges	rice leafminer
mosquitoes*	tadpole shrimp	leafhoppers
	crayfish	grasshoppers
	armyworms	

*Not a pest of rice, but populations must be controlled every year.

Major damage from many invertebrate pests can often be limited by careful management practices. Using an adequate seeding rate compensates for potential damage by stand establishment pests. Seeding immediately after flooding helps avoid the buildup of pests such as rice seed midges. Manipulating water depths can discourage feeding by some invertebrate pests. In dry-seeded rice, tadpole shrimp and rice seed midges are not a problem; rice water weevil is less of a problem. Cultural control methods also reduce pest injury. Early and effective weed control, especially of broadleaf weeds, is important in the management of several invertebrate pests.

Using pesticides is necessary in some cases. Properly timing all applications is critical for the effective and efficient use of pesticides in an IPM program. Be sure that the materials you choose to control invertebrates are compatible with herbicides. When draining water from a field treated with a pesticide, be sure that water from the field is not released before the time specified on the label for that pesticide. For specific information on treatment thresholds and pesticides, consult *UC IPM Pest Management Guidelines: Rice*, listed in the suggested reading section.

The guidelines for stand establishment pests (tadpole shrimp, midges, and crayfish) are based on a seeding rate of 150 pounds per acre (168 kg per ha). If all the seeds in a field planted at this rate germinated and established, there would be about 40 to 60 plants per square foot. However, about one-half to one-third of the seed sown is consistently lost to invertebrate pests and other factors, including seedling diseases, seed drift or burial, toxic gases, birds, or rodents. For most rice cultivars, 12 to 15 plants per square foot is considered adequate for an acceptable yield, but 20 plants per square foot is desirable for uniform grain maturity and maximum yield. The 150-pounds-per-acre rate is recommended to achieve this density. Maintaining a stand of 20 plants per square foot is also important to control weeds, but excessive stand densities encourage stem rot dispersal, so some growers may choose to use a lower seeding rate. If you are using a seeding rate that is significantly below 150 pounds per acre, the recommended threshold levels are not applicable and will have to be adjusted.

Monitoring Invertebrates

Check your fields regularly for invertebrates or evidence of their presence. Base your decisions to take corrective action on the *UC IPM Pest Management Guidelines: Rice*. The *Guidelines* will have current monitoring steps and forms for the major pests.

Check for weeds, invertebrates, and nutritional deficiencies during the same visit. Keep written records of your monitoring results because well-maintained records will help you make effective pest control decisions and predict future pest management needs.

The equipment you will need for monitoring includes a

- hand lens
- metal or plastic cylinder enclosing a 1-square-foot area (with a diameter of approximately 13.5 in, or 34 cm)
- similarly sized ring of floating plastic tubing
- glass jar 5 to 6 inches (12–15 cm) deep
- fish seine
- pencil and sampling forms or paper for record keeping

To carry out an effective monitoring program, you will have to move out into the flooded field and get wet, so wear suitable clothing.

When you are sampling, be aware that each basin within a field can be different because water levels influence invertebrate populations. Additionally, you may need to take your samples in one or two diagonal lines transecting the field (Fig. 12) to be sure you get a representative sampling of field conditions. Always move against the wind when observing

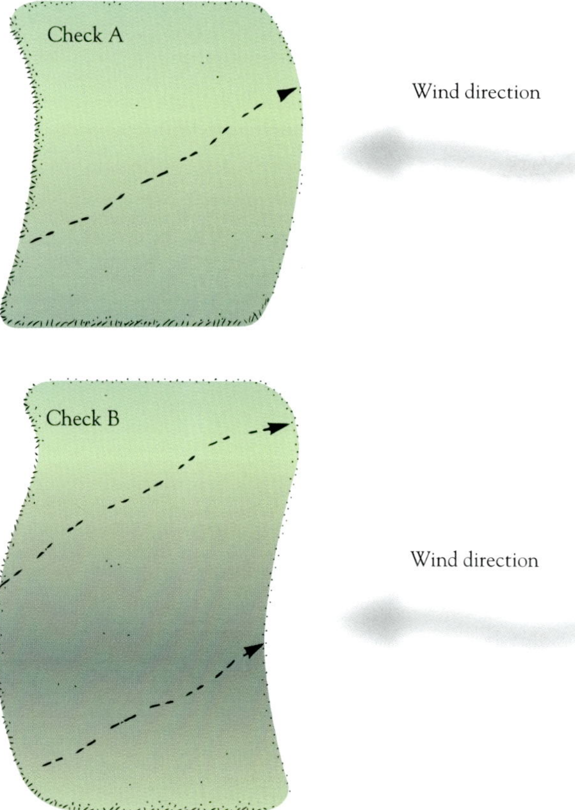

Figure 12. Sampling plan for most invertebrate pests in rice. When sampling for tadpole shrimp, seed midges, leafminers, or armyworms in rice fields, monitor each check following a path that transects the field diagonally. In a large check (check B above), take two diagonal sample sets. When sampling for underwater pests such as tadpole shrimp and midges, be sure to move against the wind so you do not muddy the water ahead of you.

aquatic pests; otherwise, the movement of your feet will muddy the water ahead of you.

Use the pictures in this manual to help you identify many common invertebrates and their natural enemies in the rice field. Check with your farm advisor or PCA for help in identifying others.

Management

Field history is important for invertebrate pest management in rice. Check all fields but give special attention to those that had significant pest numbers or injury in previous years. Large blocks of rice grown together or fields that are repeatedly sown with rice are more likely to have serious invertebrate infestations.

Prevention is key for the management of many rice pests. Seeding soon after flooding can reduce the injury by tadpole shrimp and rice seed midge. Water levels can influence pest damage, and fallow periods between crops can reduce pest numbers. Weed management can be important in limiting some invertebrate pests such as leafhoppers, armyworms, and mosquitoes.

Correctly identify the pest causing the injury using the descriptions and photos in this manual. Some pests can cause similar damage. For example, the presence of cut or uprooted rice seedlings and muddied water may indicate an infestation of tadpole shrimp. However, you should then check for shrimp and floating shrimp skins to confirm your suspicions because seed midges and crayfish may cause similar injury.

Rice fields provide a suitable habitat for mosquitoes. While mosquitoes do not damage rice, mosquitoes are a nuisance and can carry and spread diseases such as encephalitis and West Nile virus. Growers must provide land access to mosquito and vector control districts (MVDs), also called mosquito abatement districts (MADs). If problems with mosquitoes are not corrected, growers may be assessed a fee. Alert your local MVD when reflooding drained fields.

Follow the monitoring protocols and treatment thresholds developed for many of the invertebrate pests by consulting *UC IPM Pest Management Guidelines: Rice*. Treatments include pesticides, maintaining plant health, and cultural control methods such as controlling water levels.

Tadpole Shrimp

Triops longicaudatus

Although they are crustaceans, tadpole shrimp resemble tadpoles in size, shape, color, and mobility. They commonly inhabit temporary freshwater pools in low, noncultivated areas. However, when they are present in large numbers in fields with young submerged seedlings, their feeding and digging activities can reduce stands. Tadpole shrimp are not a problem in dry-seeded rice fields.

Description and Seasonal Development

Tadpole shrimp have about 35 body segments, all but the last 6 or 7 with pairs of leaflike, gill-bearing appendages. A thin, olive brown shield (the carapace) covers the front part, and two long tails (the cercopods) extend from the last segment. The true antennae are small and inconspicuous, but two longer, jointed processes resembling antennae extend from below the well-developed mandibles.

If the water is clear, you can often see tadpole shrimp swimming at the bottom of the basin.

Tadpole shrimp adults may vary in size from 1½ to nearly 3 inches.

Tadpole shrimp may chew the tips of shoots (left, top) as they emerge from the germinating seed.

TADPOLE SHRIMP

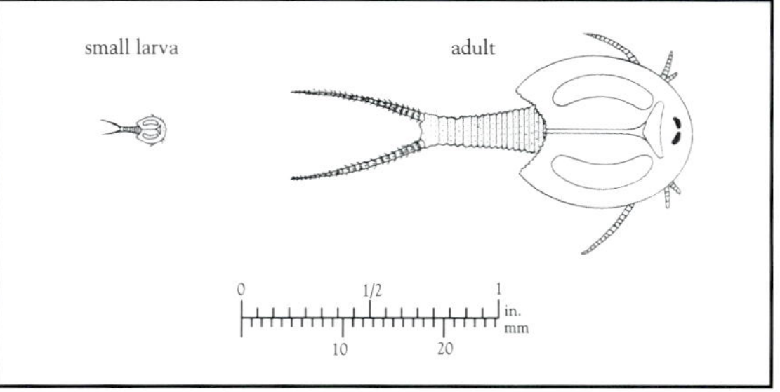

Adults are hermaphrodites (i.e., they possess both male and female reproductive organs) and carry masses of orange eggs in two brood pouches on the appendages of the 11th segment, just at the lower margin of the carapace. The adult deposits the eggs singly on soil or on plants at the bottom of the basin. Although a few eggs may develop into larvae just after laying, most require a period of drying to hatch, and populations develop almost entirely from eggs laid in previous years. The eggs are highly resistant to desiccation, remaining viable for several years in dry soil.

Most larvae begin hatching 1 to 3 days after spring flooding of the rice fields, but hatching may continue for 1 to 2 weeks. Larvae develop rapidly, molting and developing a carapace and tails to resemble young adults in less than 24 hours. Very small shrimp (¼ in, or 0.6 cm) feed primarily on microorganisms, but larger shrimp (½ in, or 1.25 cm) may chew emerging shoots and roots of rice seedlings. Larvae shed their skins several times before reaching sexual maturity and continue to molt and increase in size after they are adults. Floating molt skins may sometimes be mistaken for dead shrimp.

About 9 days after hatching, tadpole shrimp begin their reproductive phase by first digging and stirring up the silt on the bottom of the basin or pond (Fig. 13) and then laying eggs in the soil or on plants. The digging and chewing activities may uproot entire seedlings or cut off leaves, causing them to rise to the surface of the water. Masses of windblown cut leaves and uprooted seedlings floating along dikes are good evidence of tadpole shrimp activity. The shrimp causing the damage may occur only in localized populations in one or more areas of the field.

Figure 13. Digging activity of a tadpole shrimp preparing to lay eggs.

Seedlings damaged by tadpole shrimp may have all or part of their roots gnawed off.

Damage

Tadpole shrimp can cause losses in seedling rice stands in two ways. First, they may chew off seedling roots and coleoptiles or uproot seedlings with their digging, killing or injuring plants and reducing the stand. Second, their digging activities muddy the water, reducing light penetration and thereby slowing the growth of submerged seedlings. Low populations of shrimp do not cause significant economic damage. Tadpole shrimp cause no injury to rice once leaves have reached the water surface and roots are well established in the soil.

Masses of uprooted seedlings are an indication that tadpole shrimp may be causing damage. Uprooted seedlings usually drift to the borders of the rice paddy.

Tadpole shrimp have chewed on the leaves of this uprooted, floating seedling. At bottom, right is the molted skin of a tadpole shrimp.

Rice Seed Midges

Cricotopus sylvestris
Paralauterborniella subcincta
***Paratanytarsus* spp.**

Midges are the most common group of insects in rice fields. Although about 30 species may be found, only 3 are known to cause significant losses. These are *Cricotopus sylvestris*, *Paralauterborniella subcincta*, and a species in the genus *Paratanytarsus*. All three can be found in nearly every rice field every year, often in large numbers, particularly *Paratanytarsus*. Economic damage by midge larvae in rice is limited to germinating seeds and very young seedlings. They are not a problem in dry-seeded rice fields.

Midge adults are small flies with long legs and body.

Description and Seasonal Development

Adult midges swarm in the air over rice fields, levees, and around other bodies of water. Midges resemble small mosquitoes, but the mouthparts of midges are undeveloped, and they lack the scales on wing margins and wing veins characteristic of mosquitoes.

Midges prefer to deposit their eggs on open water. Masses of eggs are laid on the water surface in strings held together by a sticky, mucuslike material that swells in contact with water to form a protective gelatinous envelope around the eggs. Eggs hatch in 1 to 2 days.

Swarms of adult midges are often seen around rice fields.

The larvae use secreted silk and bits of algae or debris to build tubes on the bottom of the flooded field and on submerged vegetation. The tubes serve as retreats and also act as webs to entrap algae, diatoms, and detritus for food. Larvae may also feed on seedlings and leaves of rice and other plants. The larvae go through four instars in about 7 to 10 days in the spring when the water is warm and unshaded. The third and fourth instars cause the greatest damage to rice. Larvae pupate in silk tubes underwater.

Adults emerge at the surface in 2 to 3 days. Three or four generations occur each summer, but because damage is limited to seedlings, only the first two generations are of economic concern to rice growers.

Midge larvae live inside tubes that they build from silk and bits of debris. The larva in this picture has crawled part way out of the tube. It has already damaged the root emerging from this seed.

RICE SEED MIDGE

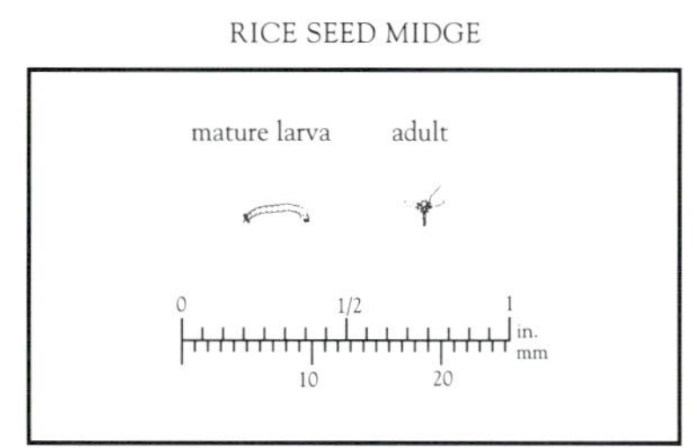

Masses of midge larvae in their wormlike tubes are often seen at the bottom of the flooded rice field.

Damage

Serious injury to rice is limited to germinating seeds and very young seedlings. Midge larvae feed on emerging leaves or roots or may hollow out embryos, killing plants. Once the seedling is 3 to 4 inches (7.6–10.2 cm) long, it can usually outgrow the feeding of midge larvae, even though irregular holes may be eaten in the leaf. Midge larvae may also feed on floating leaves, causing small holes that extend completely through the leaves. Again, injury to these older leaves is not of economic concern if other leaves are upright.

Seed and seedling injury caused by midges is most easily distinguished from injury by tadpole shrimp and other seedling pests when it is recent and the seed tissue is still firm. Midge larvae often eat the inside out of the seed, leaving it hollow; tadpole shrimp never cause this kind of damage. Sometimes damage by midges is limited to chewing away shoots and roots. Tadpole shrimp may cause similar injury, but the chewed areas on the roots, coleoptile, and young leaves are generally larger and more irregular because of the larger size of shrimp mandibles. If the damage is caused by midges, the midge larva and tube are often still on or in the seed. If the injury is several days old, secondary organisms may invade the tissue and the actual cause of damage is difficult to determine. Feeding damage caused by midge larvae on older leaves can be distinguished from damage by rice leafminer or adult rice water weevils because midges eat holes in the leaves whereas leafminers and adult weevils do not eat through the epidermis.

The three hollowed-out seeds in this picture are typical of midge damage to germinating seeds. Midges may also limit their feeding to chewing off shoots and root seedlings, leaving damage similar to that caused by tadpole shrimp (seed at top left).

Crayfish

Procambarus clarkii
Orconectes virilis

Two species of crayfish cause problems in California rice fields. The most common is *Procambarus clarkii*, but *Orconectes virilis* may also be found in canals and streams around the rice field. Both are imports from the eastern United States, where they are sometimes grown commercially for food or as laboratory animals. Unlike the native California crayfish, they are avid burrowers and can damage irrigation systems in and around flooded rice fields. Both eastern species are now widely distributed in California.

The most common crayfish in rice fields is *Procambarus clarkii*, which is red-brown and somewhat speckled.

The green-brown crayfish *Orconectes virilis* is sometimes found in canals and streams around rice fields.

Description and Seasonal Development

Crayfish reproduce once a year. Mating takes place anytime between spring and autumn. Eggs hatch in the fall or the following spring. The female carries up to several hundred eggs on the appendages in the center of her abdomen; eggs remain attached to the female until they hatch, which takes from 2 to 20 weeks, depending on the season. Young crayfish usually remain in the mother's burrow until they molt three times. After leaving the burrow, they molt six or seven more times before reaching maturity in early spring or late autumn. Adult crayfish may live up to 2 years and molt two to four more times, growing to a length of 3 to 4 inches (7.5–10 cm).

Crayfish spend most of their time hidden in burrows or sheltered areas along the bank. Their distinctive burrow holes or mounds and the damage that they cause to irrigation systems are the most obvious indicators of their presence. The burrows are filled with water, and crayfish may excavate down to a depth of 3 feet (90 cm) below the ground. Although adults rarely move around during the day unless it is cloudy, young crayfish can often be seen crawling about the bottom of the field.

Crayfish feed primarily on dead and decaying matter and plants but sometimes eat insects and other live animals inhabiting the rice field. When the field is drained, crayfish retreat to their burrows or migrate. If the burrows remain moist, crayfish can survive at least until the next season and maybe longer.

The most distinctive indicators of a crayfish population are their burrows, which are scattered along ditches, levee banks, and in the field. This photo shows burrows on the bottom of a drained field in winter.

Damage

Crayfish are of serious concern because their burrowing in ditches and levee banks may disrupt the irrigation network. Burrows near head gates and weir boxes often make it impossible to maintain an acceptable water head. Subsequent damage may be costly if the lost water contains an herbicide because it may be necessary to re-treat. Crayfish burrowing and swimming may also muddy the water, reducing photosynthesis in submerged plants. Soil forced up around burrows by crayfish after the field is drained may be picked up by harvesting machinery and contaminate harvested grain.

Crayfish eat rice seed and seedlings, and their digging may uproot seedlings as well. Although damage to seed is sometimes wrongly attributed to feeding by midges or tadpole shrimp, crayfish injury is somewhat different. Unlike these other pests, crayfish crush and macerate seed and seedlings. In a few cases, they have considerably reduced stands, but extensive injury has not been recorded as a frequent or widespread problem. Crayfish are not believed to feed on rice plants after tillering.

As it builds its burrow, the crayfish pushes lumps of mud up around the rim of the entrance.

Rice Leafminer

Hydrellia griseola

The rice leafminer is a pest of young plants and causes problems primarily where rice is grown in deep water culture and when the weather is cool. Once leaves are above water, the rice leafminer does not reduce yields even though it may still be present in the field. Rice leafminer does not cause economic damage on short-stature rice cultivars.

Crayfish burrowing may cause serious damage when it occurs near the base of the weir box because it causes leaks in the irrigation system.

Description and Seasonal Development

Leafminer adults are flies and lay their eggs singly on the upper surface of leaf blades. They prefer leaves floating on the water, and high humidity (80–100% RH) is required for hatching. Eggs hatch in 3 to 5 days. The larvae burrow into the leaf tissue, mining the leaves. Larvae may pupate in an existing mine or migrate to a different leaf to form a new mine. Total developmental time from hatched egg to adult ranges from 13 days at 90°F (32°C) to 94 days at 50°F (10°C). Larvae may be killed by extremely high temperatures (112° to 114°F [44.4° to 45.6°C]), especially when they mine upright leaves exposed to the sun.

Crayfish leave cuts on seed they damage, distinguishing them from seed damaged by tadpole shrimp and midge larvae.

Up to eleven generations may occur in a year, depending on temperature, host plant availability, and suitable high-moisture conditions. In addition to rice, a wide range of grasses associated with aquatic habitats serve as host plants. Leafminers generally overwinter as adult flies, and they may begin to lay eggs on leaves or grasses growing alongside rain pools as early as February. Up to 3 generations have been observed in rice.

Damage

Injury is caused by the larvae feeding in mines between the two epidermal layers of a leaf. The mines usually contain a swelling, which is the body of the feeding or pupating leafminer. Unlike midges, leafminers never eat holes through a leaf. Mines fade to a light green color at first, then turn yellow, and finally become white or transparent. Because high humidity is required for hatching, the leafminer attacks leaves lying on the water surface. The larvae are mobile and move on to new leaves after old ones are completely mined. In severe infestations, they may also mine the leaf sheath.

Rice leafminer adults have a metallic olive brown sheen.

Leafminers lay their eggs on floating leaves (A). The eggs are elongate and ribbed (B).

Plant vigor and weather conditions govern the extent and seriousness of the injury. Any factor that increases the number of leaves remaining prostrate on the water or the length of time they are fully in contact with water, especially deep water culture or cool weather, extends the period of time seedlings are susceptible to damage. Leaf damage at this stage reduces photosynthesis at a critical time, because food reserves from the seed have already been depleted to get the plant through the water. The plant is usually able to put forth additional leaves, but continued mining can result in reduced tillering, greater susceptibility to later pest attack, delayed maturity, or death of the plant. Once leaves start growing upright above the water, the rice leafminer cannot cause damage of economic significance.

Biological Control. Several parasitic wasps attack the rice leafminer. The most effective are believed to be *Chorebus aquaticus* and *Opius hydrelliae*. Parasitism of leafminer generations that feed on grasses prior to flooding the rice fields can be high, up to 50%. In rice, parasitism of the first generation of leafminers is low but increases to 70 to 80% during the second and third generations. These parasites lay their eggs in leafminer larvae within the leaf mines. Parasite larvae feed, grow, and mature within the leafminer larva and pupa, eventually killing it. The best way to detect parasites in your field is to check old leafminer pupae for the small holes through which the adult parasites emerged. Normally, a combination of parasites, predators, and high temperatures causes leafminer populations to drop rapidly by summer.

Rice Water Weevil

Lissorhoptrus oryzophilus

Rice water weevil larvae feed on the roots of young and mature rice plants, causing lost yield by reducing growth, tillering, and plant vigor. The weevil is a more significant pest in the upper part of the Sacramento Valley than elsewhere because of the repeated annual planting of rice, a greater uninterrupted expanse of rice fields, and possibly the weevil's longer presence in this area. However, the rice water weevil is also found in San Joaquin, Stanislaus, and Merced counties.

Leafminers feed between the leaf tissues, creating an elongated swelling. These bumps make the larger leafminer instars easy to detect in the field because you can feel them as you pass the leaves between your fingers.

Here the upper leaf surface has been removed to expose the leafminer larva. It is a typical maggot with black mouth hooks at its head end.

As the leafminer's feeding continues, the mines fade, turn yellow, and finally become transparent. The pupa can be seen underneath the scarred tissue.

Leafminers pupate within the leaf tissue. Part of the leaf has been removed here to expose the pupa.

RICE LEAFMINER

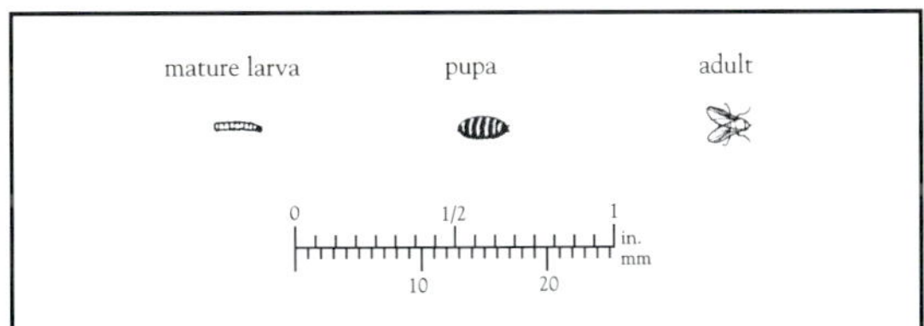

Rice plants are vulnerable to damage by leafminers only from the time the leaves begin to float on top of the water until they begin to grow upright above the water surface. This field is at the most susceptible stage, and the water is quite deep, making it even more susceptible to damage.

Description and Seasonal Development

When daytime temperatures rise above 70°F (21°C) in the spring, rice water weevil adults begin emerging from their overwintering sites at the base of grass clumps on levees and ditch banks. They feed on these grasses to build up their wing muscles and then fly anytime between dusk and midnight when evenings are warm. Flights start in March and peak in early May. Adult weevils are attracted to rice fields soon after flooding and become very active on the rice plants around the time that leaves emerge through the water. The adults feed on leaves, leaving longitudinal scars. No males are found in California, and females reproduce without mating. They lay their eggs singly underwater in the leaf sheath tissue above the plant crown. One female may lay over 200 tiny (1⁄28 in, or less than 1 mm) elongated eggs over a period of several weeks. They prefer to lay most of their eggs in parts of the basin adjacent to levee margins. Weevil larvae remain concentrated in these areas.

After hatching from the eggs, larvae mine the leaf sheath for about a day and then move to the soil to feed on the roots, where they stay through four larval instars. The legless larvae are milky white with light brown heads. When their growth is completed, larvae pupate in mud-coated cocoons, which they attach to the roots of rice, sedges, or various grasses. The silk cocoon inside the mud ball is white.

Adults emerge from the pupal cells from July until September. They feed on rice leaves, but by this time most plants are growing vigorously, and so yields are not affected by adult weevil feeding. A few of these adults lay eggs in July or August, but most enter a resting stage called diapause and overwinter in debris or at the base of plants, particularly perennial grasses.

The life cycle from egg to adult takes about 78 days in the laboratory at 73°F (22°C), with approximately 7 days in the egg stage, 50 days as a larva, and 21 days for pupation. The minimum time for development in the field is about 60 to 65 days.

Damage

Root pruning by larvae is the major cause of reduced yields. Plants with damaged roots may become stunted and lose yield through reduction of tillers and panicles or because maturity is delayed. This injury and plant yellowing that may occur can look similar to nutrient deficiency or toxicity. Reduced tillering and slower growth also allow weeds to become better established. The heaviest infestations and most serious damage can be expected to occur between late May and July within 15 to 20 feet (4.6–6.1 m) of the margins of the fields and levees, where weevils are concentrated.

Adult feeding causes linear slits of varying length on the upper surface of the leaves but generally does not cause economic losses. High populations of adults feeding on thin stands of young rice seedlings just as they emerge through the water may kill the seedlings, but such damage is uncommon.

The adult weevil spends the winter inactive at the base of bunch grasses along levees and field borders.

The rice water weevil adult has a prominent beak and dark markings on its back. Underwater, it may look darker and have a green cast.

Adult feeding causes longitudinal scars on the upper surface of the leaves. The darker scars are fresh injuries. The lighter ones are several days old.

White, sausage-shaped eggs are laid in or under the leaf sheath tissue above the crown but underwater. The eggs are hard to spot; there are two in this photo.

The weevil spends its immature stages underwater attached to rice roots. The round mud ball at the top contains the pupa.

The bumps on this larva are not legs but dorsal hooks with which the larva penetrates roots to obtain air. Note the white tube along the dorsal side of the larva for air distribution throughout the body.

These chewed-off and hollowed-out roots are typical of injury from rice water weevil larvae.

Rice water weevil damage is usually confined to the edge of fields. Broadleaf weeds have invaded the damaged areas.

Plants that have been severely damaged by weevil larvae are small, light colored, weak, and slow growing. A healthy plant is shown on the left. *Photo:* Albert A. Grigarick.

RICE WATER WEEVIL

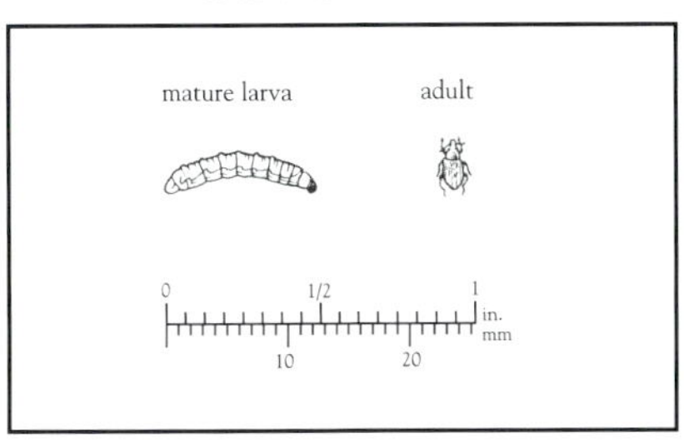

Leafhoppers

Macrosteles fascifrons

A number of leafhopper species feed on rice plants by sucking up plant fluids through their long, piercing mouthparts. The aster leafhopper, *Macrosteles fascifrons*, is the one most commonly found on rice in California. In other parts of the world, leafhoppers are significant pests because of their role in the transmission of a number of pathogens that cause serious rice diseases. Although they are not known to act as vectors of any rice pathogens in California, leafhoppers may occasionally occur in sufficient numbers to cause damage by their feeding. Injury symptoms associated with leafhoppers include stippling, yellowing, and drying leaves. Leafhoppers prefer senescent leaves, and symptoms usually occur on older leaves first.

Leafhoppers usually pass the winter in the egg stage, although nymphs and adults may be found all year round. The leafhopper inserts its eggs into tender plant tissues. Wingless nymphs hatch from the eggs and go through four or five molts before reaching maturity. Up to six generations may be completed between spring and fall.

Although leafhoppers can be present in fields during most of the growing season, the heaviest populations usually occur from early July through mid-August. Leafhoppers are very mobile; adults fly and nymphs jump. Thus infestations are rarely localized but appear generally throughout the field. High numbers of aster leafhoppers are frequently associated with basins heavily infested with broadleaf weeds and sedges. An early and effective weed control program is an important way to discourage the development of economically damaging populations of leafhoppers on weeds.

Biological Control. Predation by the spider *Pardosa ramulosa* can significantly reduce populations of aster leafhoppers.

Aster leafhopper adults have strongly veined, nearly transparent wing covers.

Nymphs of the aster leafhopper have small wing pads in their last stage and range in color from yellow to dark green.

Leafhopper feeding may cause leaves to turn yellow to brown. These nymphs are the dark form of the aster leafhopper.

Fields with high leafhopper populations may have large yellow areas, as in the background of this picture.

Armyworm

Pseudaletia unipuncta

Western Yellowstriped Armyworm

Spodoptera praefica

Rice is not an ideal host for either of these armyworms. However, both species may occasionally invade rice fields in midsummer. Western yellowstriped armyworms commonly build up in adjoining alfalfa fields or in broadleaf weeds within or around the rice field and are not known to reproduce on rice plants. The armyworm feeds and lays its eggs on a variety of grasses and grains, including rice; it is generally the fourth or fifth generation that feeds on rice, probably after other grains or grasses have dried up or are harvested.

Description and Seasonal Development

Adult moths of both species have a wing span of about 1½ inches (35–45 mm). The western yellowstriped armyworm moth has mottled forewings and silver and gray hind wings. The armyworm adult has a single white spot in the middle of its buff-colored forewing. Both moths fly at night and are often caught in light traps.

Larvae of the armyworm are active only at night or on cloudy days. However, the western yellowstriped armyworm can be seen feeding or moving about during the day as well. Although both caterpillars vary in color, you can tell them apart. Older larvae of both species have distinct yellow stripes, but the western yellowstriped armyworm has many more very thin lines along its sides and an intense black spot on the side of its first legless segment. The spot is absent on the armyworm.

Spring and early-summer generations are spent on other plants. When other food sources are depleted, larvae of either species may move into rice fields, or adult moths may fly into the rice field to lay eggs. The armyworm lays its eggs on either grasses or rice in the field, but the western yellowstriped armyworm is believed to restrict its egg laying to broadleaf weeds in the rice field.

Caterpillars develop to full size and pupate in about 3 to 4 weeks in the summer. Pupation normally takes place in the upper surface of the soil or in debris. Most mature larvae drown in the basin before reaching a suitable site. Usually only one generation a year is spent on rice.

The armyworm moth has a distinctive white spot in the middle of each forewing.

The armyworm may vary in color from almost black to green but always has a yellow stripe just below the line of prominent black spiracles.

The western yellowstriped armyworm larva has a distinct black spot above the spiracle on its first legless segment.

ARMYWORM

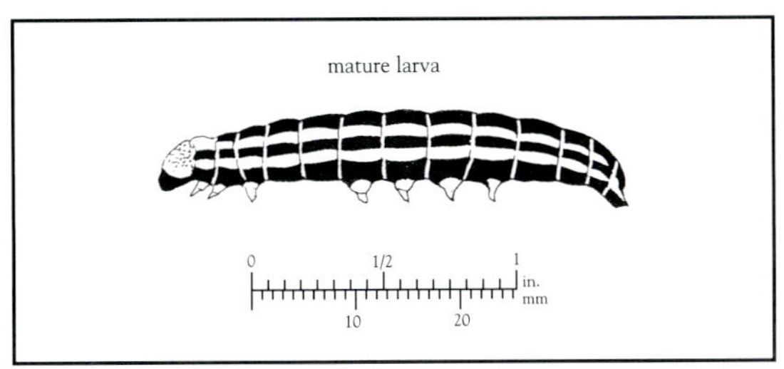

WESTERN YELLOWSTRIPED ARMYWORM

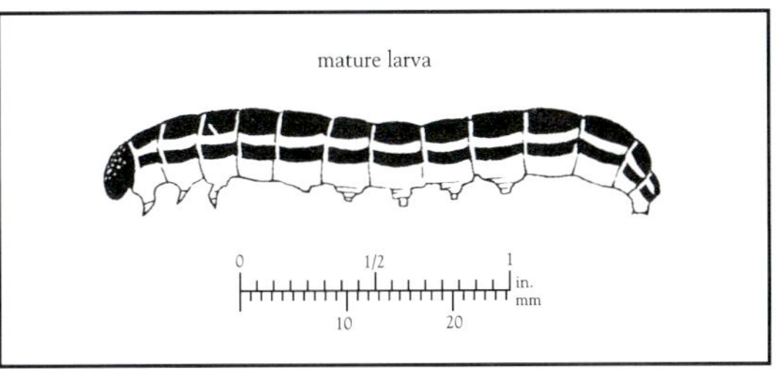

The armyworm lays its eggs under rice leaf sheaths or in folded blades (A). The blade has been pulled apart (B) to expose the eggs.

Western yellowstriped armyworm lays its eggs in large masses on various broadleaf plants. The female moth covers them with a white cottony material made of scales from her body.

Both armyworms pupate in the soil or under debris.

Damage

Injury by armyworms is most serious during periods of stem elongation and grain formation. Larvae defoliate plants by chewing angular pieces off leaves. They may also feed on the panicle rachis near the developing kernels, causing these kernels to dry before filling. This feeding leaves all or parts of the panicle white. If the whole panicle is white, the injury may also be due to low nighttime temperatures during panicle differentiation, stem rot, or rat damage. The seriousness of armyworm injury depends on the maturity of the plant and the amount of tissue consumed. Significant yield reduction can occur when defoliation is greater than 25% 2 to 3 weeks before heading.

Biological Control. Various natural factors cause mortality of armyworms in the rice field. Many caterpillars are killed by natural enemies, including predators, pathogenic microorganisms, and parasites. Natural enemies, especially when they limit spring and early-summer generations in other crops and along field margins, often keep armyworms from becoming pests in rice. The wasp *Hyposoter exiguae* often parasitizes the western yellowstriped armyworm, and *Apanteles militaris* is the most common parasite of the armyworm. The larvae of these wasps live within the armyworms until they emerge to form white cocoons on leaves (Fig. 14). Predators include the wolf spider, *Pardosa ramulosa.*

Both armyworm species chew angular pieces from rice leaves (A). Armyworm feeding can cause part of the panicle to turn white and fail to produce kernels (B). These white florets are quickly spotted in the field.

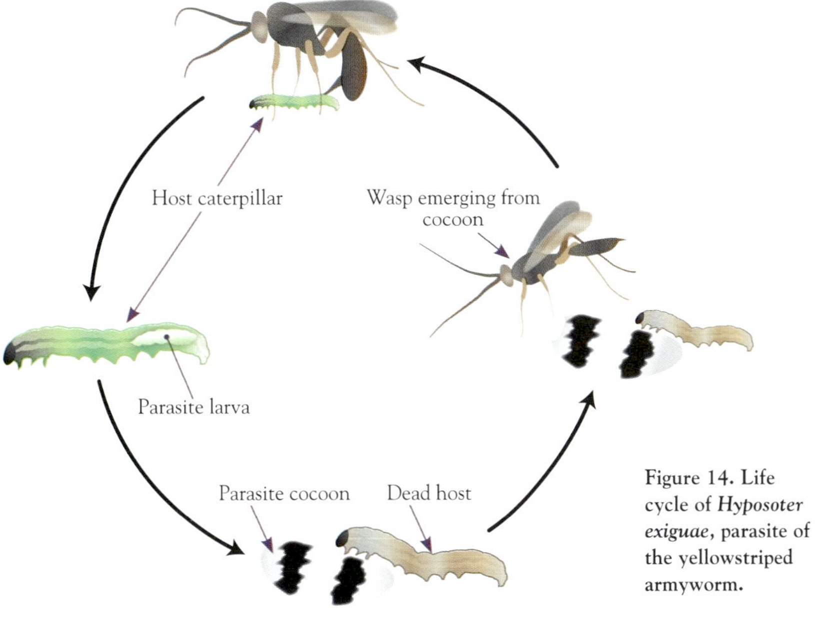

Figure 14. Life cycle of *Hyposoter exiguae*, parasite of the yellowstriped armyworm.

HYPOSOTER WASP

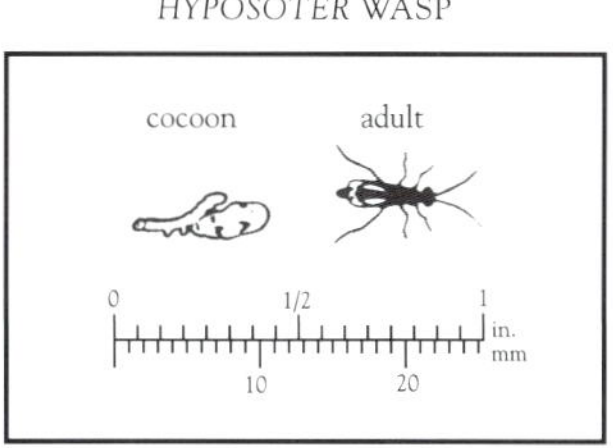

The *Hyposoter* wasp (A) lays a single egg inside the western yellowstriped armyworm. It pupates in a spotted cocoon, often attached to the dead armyworm's skin (B).

The *Apanteles militaris* wasp (A) lays many eggs inside the armyworm. Its larvae (B) develop inside the caterpillar and emerge to pupate in masses of silken cocoons on rice or grass stems.

Cocoons of *Apanteles militaris*.

Mosquitoes

Anopheles, Culex, and Ochlerotatus spp.

Rice culture provides an ideal environment for mosquito breeding. Besides being a public nuisance, certain mosquito species create public health problems. Rice fields provide habitat for the western malaria mosquito, *Anopheles freeborni*, and encephalitis mosquito, *Culex tarsalis*. *Culex tarsalis* reaches its seasonal peak in early and midsummer. *Anopheles freeborni* populations peak in late summer and fall. A third group of mosquitoes, *Ochlerotatus* species, are common during stand establishment in the spring.

Description and Seasonal Development

Adults of both *Culex* and *Anopheles* mosquitoes deposit their eggs on the water surface. *Culex* lays floating rafts of several hundred eggs. *Anopheles* eggs are laid singly. After a short incubation period, the larvae hatch. The larvae, often called wrigglers, are strictly aquatic and feed on tiny particles of plant and animal matter in the water. They do not feed on rice plants. The larvae of *Culex* mosquitoes hang vertically from the surface of the water by a breathing tube. *Anopheles* larvae breathe through a pair of spiracular plates at the rear end of their bodies and usually rest parallel to the water surface (Fig. 15). Mosquito larvae frequently wriggle down to the bottom of the basin to feed. The larvae grow through four instars and then molt into a nonfeeding pupa, or tumbler, which also hangs from the water surface. After a few days, the adult mosquito emerges from the pupa, mates, and the female flies about in search of a blood meal. During a season of high temperatures in the Central Valley, each generation takes about 10 to 14 days to complete. Numerous generations of both species are produced throughout the rice-growing season as long as the fields are flooded. *Culex* species usually reach maximum numbers in midsummer, but *Anopheles* species peak just before the fields are drained for harvesting.

Ochlerotatus mosquitoes lay their eggs on moist soil. Eggs begin to develop and hatch into larvae when the area is reflooded. Adults of this species search for other moist sites to lay their eggs. Thus, fields that are drained frequently, especially during stand establishment, may have a continuous *Ochlerotatus* problem, but in most rice fields, problems with *Ochlerotatus* are limited to the very early season.

Biological Control. Mosquito abatement districts (MADs) or mosquito and vector control districts (MVDs) combine a variety of methods to manage mosquitoes in rice fields, including stocking of fields with the mosquito-eating fish, *Gambusia affinis*. Bacteria *Bacillus thuringiensis israelensis* and *B. sphaericus* are registered for use in rice fields and used by many MADs. Several other control organisms, including fungi and nematodes that cause disease in mosquitoes or produce toxic substances, are being tested experimentally.

Many insects occurring naturally in rice fields are predators of mosquitoes. These include backswimmers, scavenger beetle larvae, giant water bugs, predaceous diving beetles and their larvae, and damselfly and dragonfly nymphs. Unfortunately, in most cases, the peak abundance of these general predators does not coincide sufficiently with that of mosquitoes to control them completely. Researchers are currently evaluating the impact of other naturally occurring invertebrate predators, including copepods.

You can limit mosquito populations by employing practices that reduce the desirability of your fields for mosquito breeding. The general themes behind these practices are maintaining the movement of water in the basin and eliminating areas of stagnant water around field perimeters. The following practices are fundamental to a mosquito prevention program:

- Make a provision for year-round rapid drainage of excess water into free-flowing main drains that are clear of vegetation.
- Construct access roads around each field for checking and repairing levee breaks and for mosquito abatement personnel to use to check fields.
- Drain and eliminate borrow pits and seepage areas external to the field.
- Cover all crop residues and green manure crops to a soil depth of 4 to 6 inches (10–15 cm) when preparing seedbeds. Large amounts of decomposing organic matter at the soil surface provide an especially attractive site for mosquitoes to lay eggs. Covering residues also reduces the likelihood of scum, or algae, buildup, which is also attractive to mosquitoes.
- Practice good weed control within the rice field and along the levees. Mosquitoes retreat to weedy, swampy areas; as adults, they use these areas to lay eggs; as larvae, they are protected in these areas from wind and predators.
- Seek the assistance of staff from the local mosquito abatement district (also called mosquito and vector control district) to develop the best possible abatement program for your operation. Alert the abatement district when reflooding drained fields.
- Several pesticide materials are toxic to mosquito-eating fish. Check with your mosquito abatement district to find out how to minimize impacts of pesticides used in your pest control program.

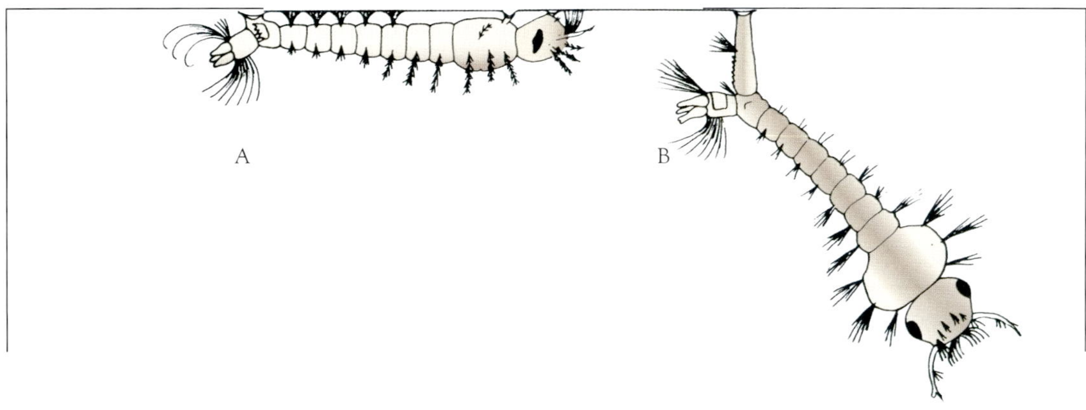

Figure 15. Resting positions of mosquito larvae: (A) *Anopheles*, (B) *Culex*.

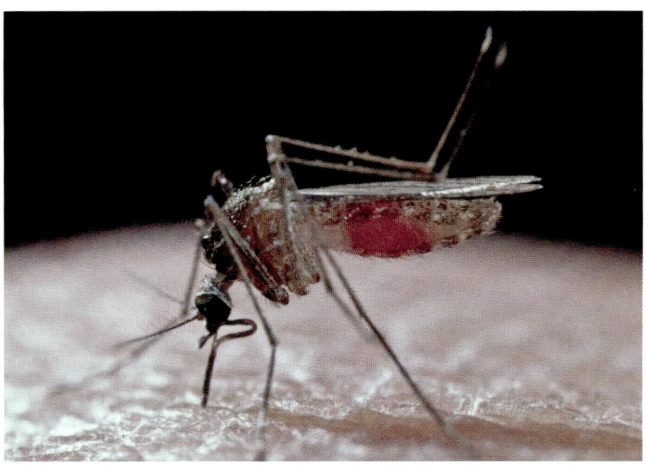

Culex tarsalis is a carrier of western equine encephalomyelitis and St. Louis encephalitis.

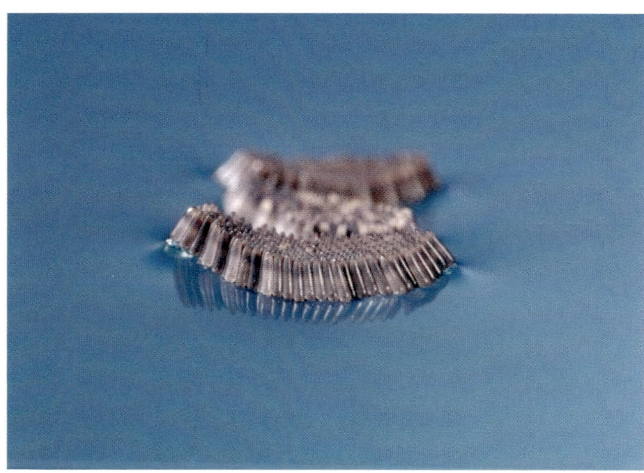

Eggs of *Culex* mosquitoes are laid in floating rafts of several hundred eggs.

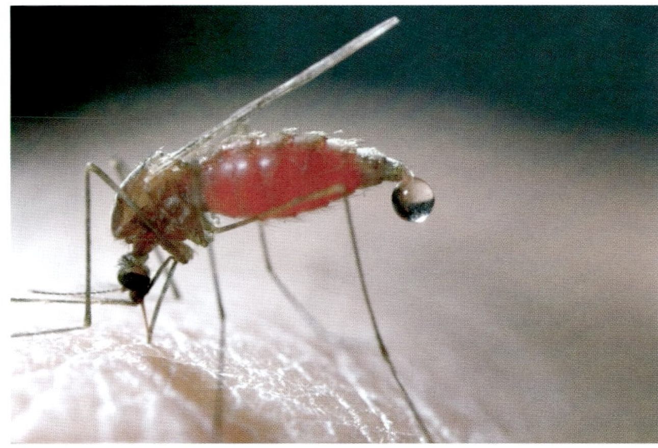

Anopheles mosquitoes can spread malaria.

Culex larvae hang from the water surface by their breathing tubes.

The fish *Gambusia affinis* is an important predator of mosquitoes.

Backswimmer adult.

Scavenger beetle larva. *Photo:* L. Dunning.

Damselfly nymph. *Photo:* L. Dunning.

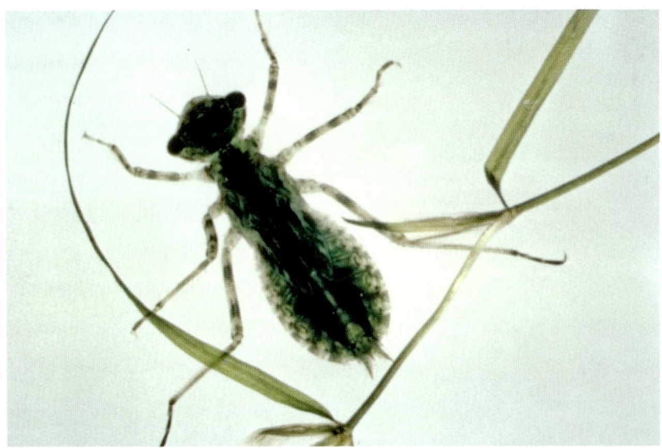

Dragonfly nymph. *Photo:* L. Dunning.

Provide road access around your fields to check for leaky levees and to allow easy access for mosquito abatement district personnel. *Photo:* Milton D. Miller.

Other Invertebrates in Rice Fields

The rice field's swamplike environment provides a good habitat for a variety of invertebrates. Besides insects, these include animals ranging in complexity from protozoans and segmented and nonsegmented worms to sizable crustaceans and snails. These organisms generally have life cycles well suited to rice culture, featuring eggs or other inactive stages that can survive while the field is drained. Most cause little or no crop damage even though they may occur in extremely large numbers. Some species are predaceous and beneficial. Don't treat for any unknown invertebrate without checking with your farm advisor or other expert to confirm the potential for damage. A few of the more common minor invertebrates are pictured and described below. Chemical treatments are generally not recommended for any of them.

Large Water Scavenger Beetle

The large water scavenger beetle, *Hydrophilus triangularis*, is often observed in flooded rice fields. Although adults may cut rice leaves to form a float for their eggs and occasionally dislodge seedlings with their swimming activities, their numbers are not great enough to cause economic damage. Adults feed on algae. The larvae, which are several inches long when mature, are predaceous on a range of other aquatic organisms, notably mosquitoes.

Rice Leaffolder

The rice leaffolder, *Lerodea eufala*, feeds on rice leaves, chewing angular holes out of the sides of the blades. The caterpillar is a dull light green with three light-colored lines running down the side of its body. A single white band runs below the spiracles, and there are two cream-colored ones above. The leaffolder pupa is green and hangs from rice plants by a silken thread. Populations have not been recorded at damaging levels, and control measures are not recommended.

Adults of the large scavenger beetle are shiny, black, and about 1.5 inches (3.8 cm) long.

The silken egg case of the large scavenger beetle may be found either floating free or attached to a cut rice leaf.

The rice leaffolder is sometimes found in rice fields but is not known to cause economic damage. The rice leaf in this picture has been unrolled to reveal the green leaffolder caterpillar inside.

Thrips

You can sometimes find various thrips species feeding on rice plants along the margins of the field. The presence of these tiny insects is usually associated with the maturing and drying of a different, adjacent crop. Thrips feeding may cause yellowing and stippling of leaves and may at first be mistaken for leafhopper damage. Check yellowed leaves for thrips with a hand lens. Chemical treatment for thrips is not recommended.

Grasshoppers

A number of species of grasshoppers may occasionally feed on rice. They chew angular holes in leaves, causing an injury similar to that caused by leaffolders or armyworms. Damage may occur throughout the field, but it is usually most prevalent around field margins. Chemical treatment is generally not necessary.

Aphids

Several species of aphids may be found on rice. They often appear after an herbicide treatment drives them off other plants. Aphids have not been known to reach levels on rice that require chemical treatment in California. In fact, studies show that aphids on weedy ducksalad significantly reduce the size of the plant and its seedpods during the latter part of the rice-growing season.

This tiny insect is a thrips. You will need a hand lens to see it clearly.

Grasshoppers are occasionally found in rice fields but rarely cause significant damage other than along field margins.

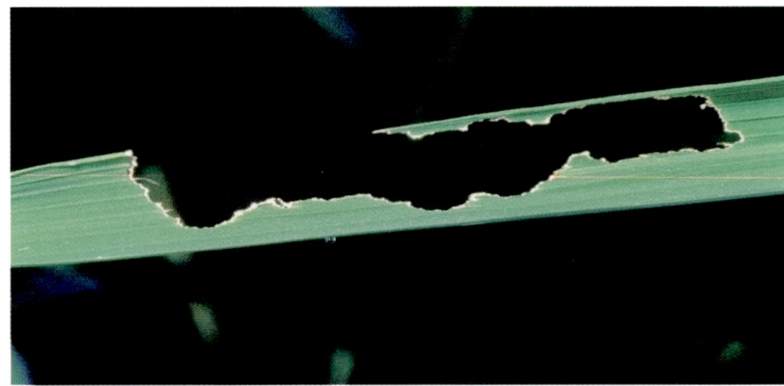

Grasshopper damage can be confused with that caused by armyworms, but grasshoppers tend to leave more of the leaf margin intact, chewing out the center of the leaf.

Although aphids may occur in rice, treatment is not recommended.

Insects Infesting Stored Rough Rice

A number of insects may infest stored rice. Their feeding causes direct damage to the rice kernels, and their activities add moisture and heat to local areas in the rice bin, often causing a loss in grain quality and enhancing secondary fungal infections. Inspect stored rice regularly for insect infestations.

Primary grain pests attack and damage whole kernels. These pests include the lesser grain borer *(Rhyzopertha dominica)*, the Angoumois grain moth *(Sitotroga cerealella)*, and the rice weevil *(Sitophilus oryzae)*. Secondary pests feed on the surface of the kernels or attack broken or previously infested kernels. The cadelle *(Tenebroides mauritanicus)*, the saw-toothed grain beetle *(Oryzaephilus surinamensis)*, the flat grain beetle *(Cryptolestes pusillus)*, and the Indian meal moth *(Plodia interpunctella)* are the most commonly found secondary pests.

For descriptions and photographs of stored product pests, see the USDA Agricultural Handbook No. 500, *Stored Grain Insects*, available from the U.S. Government Printing Office, or *Rice Quality Handbook*, listed in the suggested reading.

Management Guidelines

Prevent infestations by practicing good drying and sanitation practices. Dry grain is less likely to be invaded, and clean storage areas are less likely to harbor insects to start an initial infection. Clear storage bins out, treat them and the interior or exterior of buildings if necessary, and remove all weeds and other trash around storage areas.

Inspect stored grain regularly for insects. Chemical treatments can be added to the grain as it goes into storage or used in airtight structures after rice is in storage. Consult your local agricultural commissioner or farm advisor for current pesticide recommendations for stored rice.

Photo: Mike Poe

Diseases

Compared to many of the world's other rice-growing regions, California has a limited spectrum of rice disease problems. However, several fungal diseases can cause substantial yield losses. No bacteria or viruses are known to cause losses in California rice.

Seed rot and seedling diseases, stem rot, aggregate sheath spot, and bakanae are the most common rice diseases in California. All can cause serious economic losses if not properly managed. Rice blast was first found in California in 1996 and can cause serious problems when the pathogen is present and conditions are favorable for disease development. Kernel smut occurs on all cultivars but is most prevalent on long-grain ones. False smut, brown spot, and Pythium crown rot occur but usually do not cause economic losses.

Environmental factors, nutrient deficiencies, or toxicities can also cause production losses, often producing characteristic symptoms on the rice plant. These abiotic disorders can kill rice plants or seriously limit yields. The best known is zinc deficiency, often erroneously called alkali disease. Other abiotic disorders encountered in California are caused by toxic gases and organic acids produced when a large amount of soil-incorporated organic matter decomposes in the flooded environment. Deficiencies of zinc, iron, nitrogen, phosphorus, and potassium or toxic concentrations of boron, salts, or herbicides may occur in various situations or production areas.

Management

Good cultural practices play a major role in managing rice diseases. Removing rice straw after harvest or incorporating the straw followed by winter flooding greatly reduce the inoculum of major disease pathogens, such as those that cause stem rot and aggregate sheath spot. Avoiding excess levels of nitrogen fertilizer, especially by using split applications, and providing adequate levels of potassium reduce the severity of stem rot, aggregate sheath spot, and rice blast. Using practices that promote rapid germination and seedling growth reduces the incidence and severity of seed rot and seedling disease. Using clean seed is an important part of managing rice blast, and seed treatment is the primary means of controlling bakanae. Planting less-susceptible cultivars is an option for managing several diseases. For detailed management information on rice diseases, consult the latest editions of the *California Rice Production Workshop Workbook* and *UC IPM Pest Management Guidelines: Rice*. These publications are listed in the suggested reading at the end of this book.

Seed Rot and Seedling Disease

***Achlya klebsiana, Pythium* spp.**

Rice seedling diseases, caused by water molds *Achlya klebsiana* and *Pythium* species, can limit establishment of uniform stands of water-sown rice. These diseases occur wherever rice is water-seeded but are most severe when cool temperatures following planting retard seedling development.

Symptoms and Damage

Seed rot and seedling disease symptoms appear within a few days after planting. The most common sign is whitish hyphae growing over the surface of the seed or the very young seedling. Hyphae originate from cracks in the glumes on the seed and within a few days grow to form a halo of mycelium radiating from the infection point. You can often discern the sporangia of *Achlya*, which appear as swellings at the end of some strands of mycelium. Algae soon colonize the halo, turning it green. In some cases, a dark circular spot appears on the soil surface around the infected seed. This spot may also be due to algae, but probably it is more commonly caused by secondary invasion of the infected seed and fungus by various aquatic bacteria.

If seed rot organisms infect the plant after some primary leaves and roots have formed, these leaves and roots usually become stunted. The leaves and leaf sheaths become discolored and further development is retarded; infected seedlings also display the typical halo of mycelium. Once seedlings are established, seed rot and seedling disease fungi have little effect on plant growth and survival.

Seasonal Development

The pathogens that cause seed rot and seedling diseases survive in the soil and produce motile spores called zoospores when soil is flooded. Zoospores are attracted to exudates leaking from cracks in the rice seed coat and to germinating seedlings.

A halo of mycelium grows over the surface of seed infected with seed rot and seedling disease pathogens. The greenish cast is caused by algae growing over the mycelium.

Other microorganisms colonize the mycelium and rotting seed, often causing a black spot to appear on the soil beneath infected seeds.

The bright white elongated swellings at the tips of some of these mycelial strands are sporangia of *Achlya klebsiana*.

Stem Rot

Magnaporthe salvinii (Sclerotium oryzae)

Stem rot is the most serious of the widespread rice diseases in California. All California rice cultivars are susceptible to the stem rot fungus, although some cultivars exhibit some tolerance and differ in the severity of disease sustained.

Symptoms and Damage

The first sign of the disease in the field is the appearance of small dark lesions on rice leaf sheaths at the water level. These lesions increase in size, enlarging up and down the leaf and toward the center of the culm as the season progresses. The outer infected sheaths eventually die and slough off. When the infection spreads into the culm, the inner tissue turns dark brown to black and mycelia and sclerotia often become visible. Culm infections reduce both grain quality and panicle size. Infections very early in the season kill tillers or inhibit panicle formation. Lodging of infected plants can cause additional losses. Losses are most severe when infections begin early in the season; however, later infections contribute greatly to the increase of inoculum and thus to the potential for losses in future crops.

The first signs of a stem rot infection are small dark lesions (arrow) on rice leaf sheaths at water level.

Leaf sheaths with stem rot lesions eventually die and slough off.

White patches of mycelium are present on leaf sheaths severely infected with the stem rot pathogen.

Dark mycelial strands of the stem rot pathogen are growing in toward the culm on the leaf sheath at the bottom. The dying leaf sheath, top, is about to slough off.

Severe stem rot infection, especially early in the season, can kill emerging panicles.

Hard, black sclerotia of the stem rot pathogen *Sclerotium oryzae* develop on infected plant tissue from the time when plants begin to mature until after harvest.

The stem rot pathogen overwinters in rice residues remaining in rice fields after harvest.

Sclerotia of the stem rot pathogen float to the water surface and infect rice plants at the waterline.

Seasonal Development

The stem rot pathogen *Sclerotium oryzae* is the sclerotial stage of *Magnaporthe salvinii*. This fungus overwinters from one season to the next as a compact mass of hyphae in a hard, black, resting body called a sclerotium. Sclerotia can survive free in the soil but are more often associated with rice plant residues remaining in the field after harvest. The following season, after the field is flooded, sclerotia float to the water surface and infect leaf sheaths of young rice plants at the waterline. Sclerotia continue to rise to the water surface throughout the growing season. Infection spreads up and down the leaf sheath and into the culm, and if severe, ultimately kills the infected tiller.

The fungus forms both sexual and asexual spores during the growing season (Fig. 16), but sclerotia are believed to be the primary source of pathogen dispersal and inoculum for infection. Sclerotia begin to form abundantly on infected tissue as the plant approaches maturity. The fungus continues to grow and form sclerotia on rice residue for as long after harvest as moisture and temperatures remain favorable.

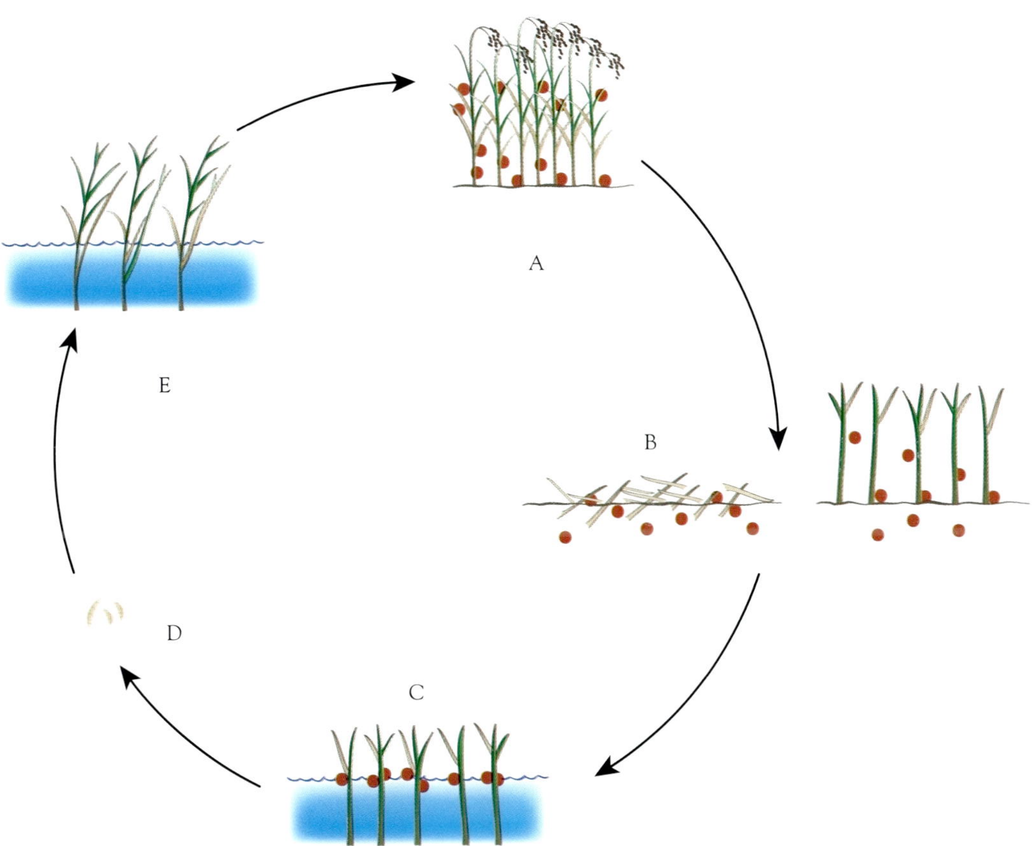

Figure 16. Disease cycle of stem rot in rice. Sclerotia begin to form on maturing plants in late summer (A). They overwinter in stubble in the ground (B) and float to the water surface to infect the next year's rice crop after the fields are flooded in the spring (C). In the summertime, the fungus forms sexual and asexual spores (D), which may infect the crop (E), but this means of dispersal is not believed to be as important in disease spread as sclerotia.

Aggregate Sheath Spot

***Ceratobasidium oryzae-sativae* (*Rhizoctonia oryzae-sativae*)**

Although the symptoms it causes are similar, the pathogen causing aggregate sheath spot in California is different from the *Rhizoctonia solani* fungus that causes the sheath blight disease responsible for serious losses in the southern rice-producing states. Aggregate sheath spot has become more prevalent in California with the increased use of short-stature cultivars, and has caused minor to severe damage in some fields.

Symptoms and Damage

Aggregate sheath spot first appears as lesions on lower leaf sheaths at the waterline during the tillering stage. Lesions appear as irregular-shaped spots with gray-green to straw-colored centers surrounded by distinct brown margins. Frequently, additional margins form around the initial lesion, producing a series of concentric bands. A strip of light brown necrotic cells runs down the lesion center. Lesions often coalesce and may cover the entire leaf sheath. Leaves of diseased sheaths turn bright yellow and then die. Although they occur at about the same location on the stem, these lesions are distinctly different from the small lesions that are the first visible signs of stem rot infection.

Seasonal Development

The disease cycle of aggregate sheath spot is very similar to that of stem rot (Fig. 17). The fungus overwinters as sclerotia in crop residue or in soil. Sclerotia float to the water surface after the rice fields are flooded. Developing from these sclerotia, the fungus infects leaf sheaths at the waterline. Aggregate sheath spot progresses rapidly at heading, spreading up the sheaths to the leaves, and under favorable conditions (high humidity or rain) can spread to the flag leaf and panicle rachises, killing entire tillers. Infected plants frequently lodge after heading. Later in the season, the fungus begins to produce new sclerotia on or in diseased tissue. These sclerotia overwinter and are the source of primary infections in subsequent rice crops. Because aggregate sheath spot is favored by warm, humid conditions, plants in dense stands are most vulnerable to infection and increased disease severity. The distribution of the disease in a field is usually spotty, although large areas of the field may be affected when conditions are favorable.

Aggregate sheath spot infections begin as large, single, gray spots at the waterline.

Many gray spots appear as the disease, aggregate sheath spot, progresses up the leaf sheath.

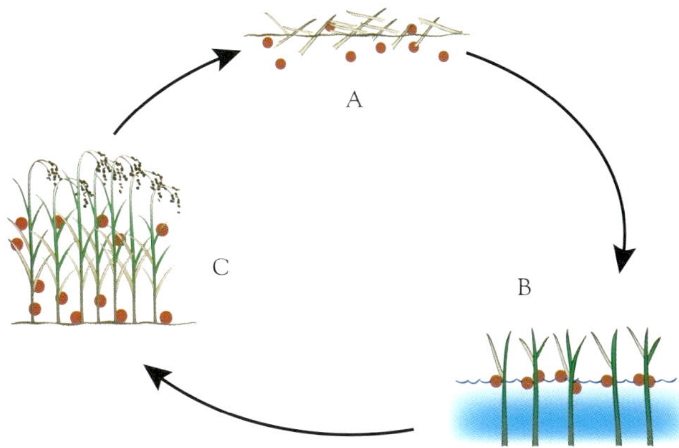

Figure 17. The aggregate sheath spot pathogen overwinters as sclerotia on crop residue and in the soil (A). In flooded fields, sclerotia float to the surface, and the pathogen infects leaf sheaths at the waterline (B). Sclerotia are produced on infected plants later in the season (C).

Bakanae

Gibberella fujikuroi (Fusarium fujikuroi)

Bakanae is a disease that occurs commonly and is very damaging in Asian rice. It was first identified in California in 1999 and has since become widespread in all growing areas, causing significant losses in some cases.

Symptoms and Damage

The most prominent symptom of bakanae is the abnormal elongation of seedlings, which are pale and prominent next to shorter healthy seedlings. Infected seedlings may also be stunted and yellow, and may develop root and crown rot. Seedlings usually are killed. If older plants are infected, they usually develop abnormal elongation, and adventitious roots may form at the lower nodes of the stems. Sometimes infected plants will be symptomless until heading, when they collapse and die.

Leaf sheaths of dying plants usually are covered with white or pink fungal mycelium and spores at the waterline. Plants that survive produce no panicles or panicles with no grain. Stems of infected plants may turn bluish black from the production of fruiting bodies (perithecia) by the pathogen.

The most prominent symptom of bakanae is the abnormal elongation of the stems of young plants, which stand out (arrows) from healthy plants.

Bakanae-affected rice plant (right) next to a healthy plant.

White or pink fungal tissue develops at the waterline on leaf sheaths of rice plants dying from bakanae.

Seasonal Development

The fungus that causes bakanae survives between rice crops primarily as spores on the surface of infested seed. It may also survive in infested crop residue but does not survive very long in the soil. As plants develop from infested seed, the fungus infects the roots and/or crown and develops systemically within the plant. Spores produced on infected plants are moved by air currents and contaminate rice seed after heading or during harvest. The disease cycle of bakanae is illustrated in Figure 18.

Production of gibberellin growth hormones by the fungus causes the abnormal stem elongation that is characteristic of this disease. Stunting of infected plants results from production of fusaric acid by the pathogen. The types of symptoms exhibited by infected plants depend on the strain of the pathogen, environmental conditions, and the nutritional status of the host plant.

Rice Blast

Magnaporthe grisea (Pyricularia grisea)

Rice blast is the most important disease of rice worldwide but was identified for the first time in California in 1996. It occurs sporadically but can cause severe crop losses if the pathogen is present and conditions are favorable for disease development.

Symptoms and Damage

Rice blast progresses through a series of symptoms as plants develop from seedlings through maturity. Leaf symptoms, or leaf blast, are the first to be seen. Diamond-shaped lesions appear on leaves. They have a white or gray center with a dark green margin that turns brown. The centers of lesions may turn bluish gray and cottony as spores are produced by the pathogen. Leaf blast may kill young plants, but leaf blast symptoms usually decline later in the season.

Leaf collar symptoms (collar rot) appear as brown or reddish brown discoloration at the junction of leaf blade and leaf sheath. Collar rot may kill entire leaves and has a pronounced effect on yield when it destroys flag leaves.

As the plant approaches maturity, stem nodes may be attacked. Infected nodes turn brown or black, and the entire stem above the node may be killed. The most destructive phase of the disease occurs when the node just below the panicle is infected (neck blast). If neck blast occurs early, it may kill the entire panicle, leaving it straw colored and blank—lacking any grain. If it occurs later, it may cause incomplete grain filling and poor grain quality. Panicle branches and spikelet pedicels may also be infected, reducing yield and grain-milling quality.

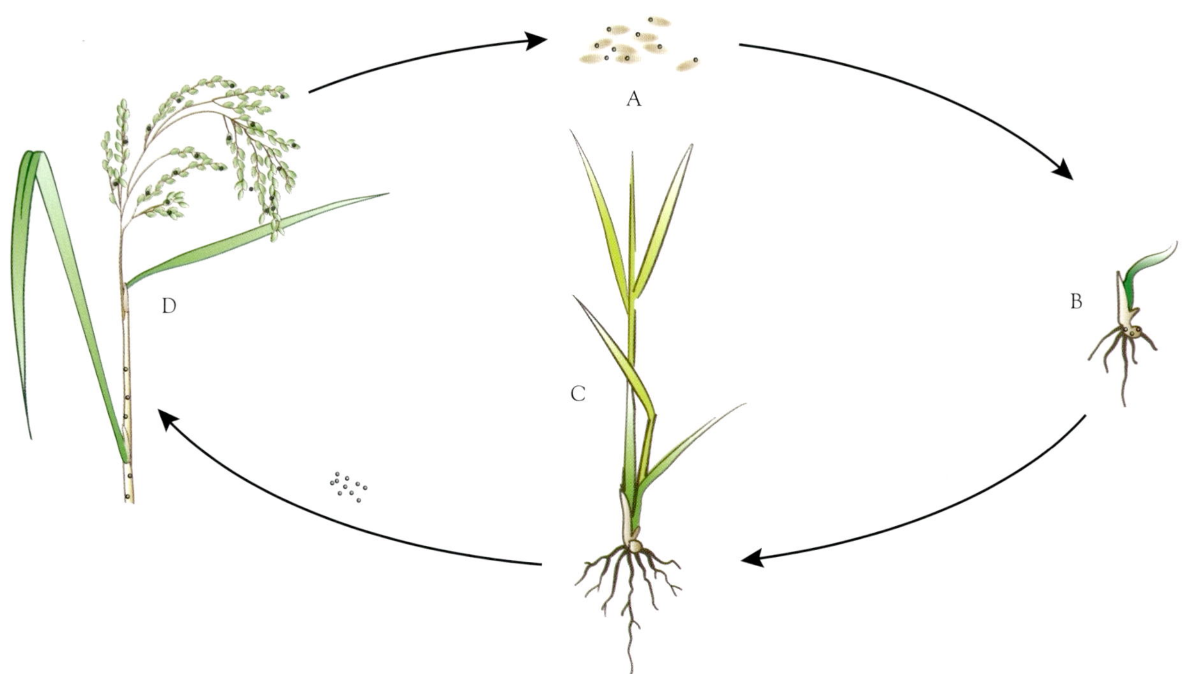

Figure 18. The bakanae pathogen survives as spores on the surface of infested seed (A). The pathogen develops systemically within plants that develop from infested seed (B). Compounds produced by the pathogen cause disease symptoms (C). Spores produced on infected plants contaminate rice seed after heading (D) or during harvest.

Newly forming leaf blast lesions have white centers and dark margins.

Leaf blast lesions are diamond shaped with gray or white centers and dark brown margins.

Rice plants may be killed by the leaf blast phase of rice blast.

The collar rot phase of rice blast appears as a brown discoloration of the leaf collar, the junction of leaf blade and leaf sheath.

Collar rot may kill the affected leaf blade.

Neck blast—the infection of the node just below the panicle—may kill the entire panicle (A). If infections occur later in the season, grain will be present in those flowers that filled before the panicle was killed (B).

Seasonal Development

The rice blast pathogen overwinters in infected seed and infected plant residue. Spores produced on infected tissue are spread by wind to the foliage of developing plants. Plant surfaces must remain wet for extended periods before infection can occur. Once infection occurs, lesions develop and new spores are produced within a week when conditions are favorable. Spores are spread to other plants, and the disease cycle (Fig. 19) may be repeated several times during a season. Therefore, the disease can spread very rapidly within a field and can spread to fields not previously infested.

Rice blast is favored by high nitrogen levels, prolonged periods of leaf wetness, high relative humidity, calm conditions, and nighttime temperatures from 63° to 73°F (17° to 23°C).

Kernel Smut

Tilletia barclayana

Kernel smut is generally a minor disease but under favorable environmental conditions may cause severe loss in some fields. It is most prevalent on long-grain cultivars but has also been observed on short- and medium-grain cultivars.

Kernel smut is first evident after heading when grain is almost mature. Kernel smut appears as a black mass of chlamydospores that replace all or part of individual kernels. Usually only a small number of kernels are infected on each affected panicle. Completely smutted kernels may be slightly swollen and may break open, exposing the black spore mass. These masses of black chlamydospores make the disease easy to recognize. Under severe disease conditions, dark spore clouds may be visible behind combines during harvest.

This rice panicle is infected with kernel smut. Note how the black chlamydospores have replaced the endosperm of the rice grains. *Photo:* Robert K. Webster.

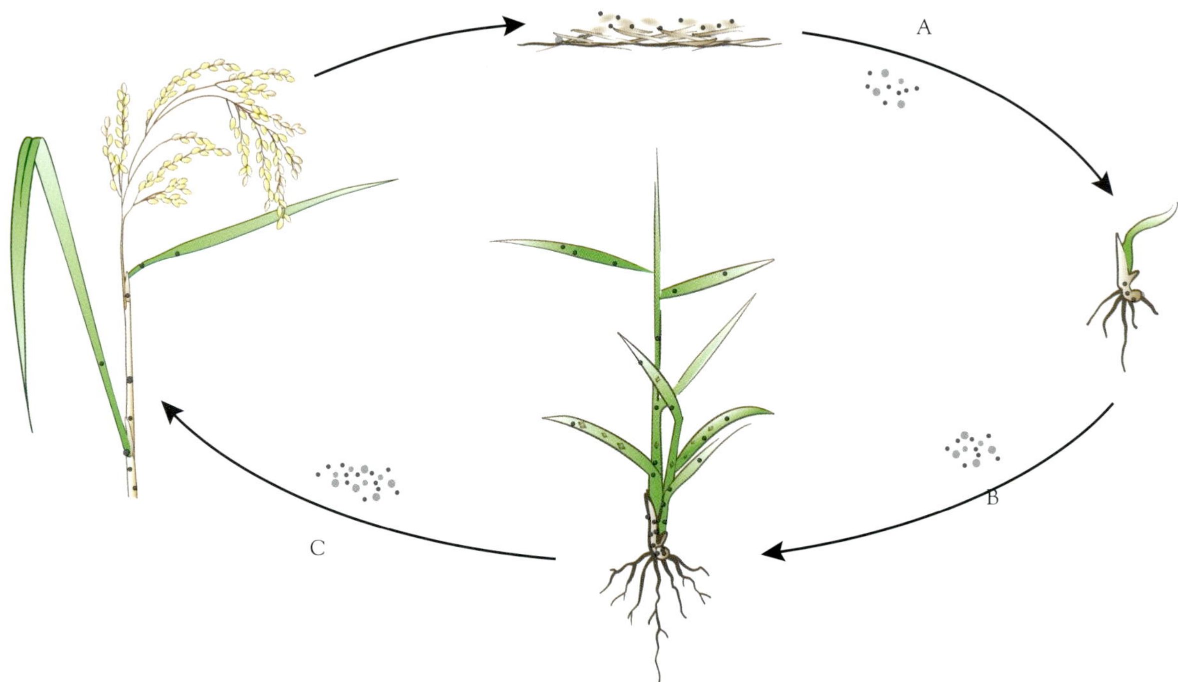

Figure 19. The rice blast pathogen overwinters in infected seed and plant residue (A). Spores produced on infected tissue infect the foliage of developing plants (B). Spores produced on these plants infect other plants, and the disease cycle is repeated when conditions are favorable (C).

Chlamydospores of *T. barclayana* may overwinter in soil and on or in seed. In spring, they float on the water, where they germinate and produce another spore stage. These spores remain dormant on the water surface until just before heading, when they germinate to form a mycelial mat. Secondary sporidia are then produced and become airborne, infecting individual florets during heading and anthesis.

False Smut

Ustilaginoidea virens

False smut, also called green smut, is a minor disease of rice in tropical areas and the southern United States. The disease was first identified in California in 2006. Presently, it occurs only in a few localized areas and its impact is unknown. No management actions are generally necessary for false smut.

The fungus that causes false smut infects developing flowers and may infect maturing grain. Grain tissue is replaced by spore balls, which are orange when immature and olive black when mature. Spore balls have a velvety appearance and, if cut open, are white inside. Sclerotia are formed in the center of spore balls. Usually only a few grains of a panicle are affected. The pathogen survives between crops as sclerotia or hardened spore balls called pseudomorphs. The disease cycle is not well understood, but it is believed that spores produced from surviving sclerotia begin the disease cycle by infecting developing flowers. Spores produced by these infections later infect maturing grain.

Brown Spot

Cochliobolus miyabeanus (Bipolaris oryzae)

Brown spot, also known as Helminthosporium brown leaf spot, has caused damage in California only in periods immediately following unseasonable July or August rains. Disease development requires free moisture on the rice leaf surface. Generally the relative humidity in the plant canopy during the summer is too low in California's rice production areas to provide these conditions; however, the disease is of major importance in the southern United States, particularly as a seedling disease of drill-seeded rice.

Cochliobolus miyabeanus is seedborne and develops on seedlings; however, it is not generally noticed until later in the season when leaves and heads of older plants become infected. Spots on leaves and grain are roughly circular to oval, dark brown or gray, but vary in size and color according to the rice variety. In severe infestations, leaves may die. Weak plants or plants in dense stands are most severely affected. Diseased plants produce lightweight or chalky kernels with spotted hulls and sometimes spotted kernels.

Where the disease occurs, maintain optimal conditions for rice growth to limit damage by brown spot.

ABIOTIC DISORDERS

Rice plants usually display characteristic symptoms when they have an inadequate supply of plant nutrients or are exposed to toxic levels of naturally occurring or synthetic chemicals. Symptoms caused by abiotic factors may resemble those caused by disease organisms. Correct identification of plant symptoms and their cause is necessary for an effective integrated pest management program. The most common and serious abiotic disorders include plant nutrient deficiencies, exposure to naturally occurring toxic acids, gases, or compounds, and synthetic herbicides and herbicide-insecticide interactions that injure rice.

Nutrient Deficiencies and Toxicities

Many nutrient deficiencies and toxicities cause rice plants to develop distinctive symptoms. A diagnostic key to these symptoms is presented below. Symptoms of nutrient deficiency and toxicity usually occur in irregular or mottled patterns throughout the field. Soil irregularities, such as sandy streaks, dry spots, alkaline or saline areas, exposed calcareous subsoils, uneven fertilizer application, hardpan areas, and different soil types are responsible for these patterns.

Leaf symptoms caused by deficiencies of immobile nutrient ions, such as iron, zinc, manganese, boron, and copper, usually occur in the young or upper leaves because immobile ions are not translocated from older to younger leaves. Consequently, older leaves may have normal color in the early stages of nutrient deficiencies. This feature can be confusing on plants recovering from certain deficiencies. This is especially common with zinc; after an initial deficient condition induces plant symptoms, roots of young plants often develop sufficient root mass to reach new sources of zinc. New growth then begins to appear healthy and green while old growth remains stunted and yellow or brown.

On the other hand, symptoms caused by deficiencies of mobile nutrient ions, such as nitrogen, potassium, phosphorus, or sulfur, usually occur initially in the lower leaves of the plant, with new leaves showing no symptoms. In deficient plants, these mobile nutrients are translocated from the older leaves to the younger parts of the plant, which prevents the appearance of symptoms. Late in the season, deficient nutrients are translocated from some of the upper leaves to the panicles and grain. This nutrient movement usually causes an irregular, blotchy pattern, often starting at the leaf tip or margin.

When making a diagnosis, look for plants with moderate symptoms; the symptoms on these plants are usually more definitive than on severely affected plants. On plants with extreme symptoms, tissue is usually dead or dying, and all dead leaves look very much alike. Always confirm suspected nutrient deficiencies or toxicities with a soil or leaf analysis. Methods for carrying out soil and leaf analysis and correcting nutrient deficiencies are presented in the *California Rice Production Workshop Workbook* and the *Rice Quality Handbook*, listed in the suggested reading.

Key to Nutrient Deficiencies and Toxicities

The following key is a guide to help diagnose some of the more common nutrient deficiencies and toxicities. The key is generalized; plants will not always display these symptoms, and some symptoms may occasionally be associated with other disorders. Always confirm your diagnosis with a plant tissue analysis by a reputable diagnostic laboratory. Nutrition imbalances not commonly found in California are not listed; these include manganese, copper, calcium, boron, and magnesium deficiencies and manganese, iron, and aluminum toxicities.

COLOR CHANGE IN UPPER (YOUNG) LEAVES (IMMOBILE IONS)	
1. Interveinal areas and later entire leaves lose green color and turn pale yellow to whitish with base of plants usually greener; old roots may remain white; plant may die.	IRON DEFICIENCY*
2. Seedling leaves turn yellow and tend to float on water rather than grow upright; stunted growth; brown spots and streaks later appear on lower leaves. New leaves of recovering plants may be green. Symptoms occur in patchy locations in the field; maturity delayed; plant may die.	ZINC DEFICIENCY
COLOR CHANGE IN LOWER (OLDER) LEAVES (MOBILE IONS AND TOXICITIES)	
1. Color of older leaves changes from green to light green or yellow; young leaves are greener. Leaves are narrow, short, and erect. Plants are stunted, with a limited number of tillers.	NITROGEN DEFICIENCY†
2. Leaves initially yellow but rapidly change to grayish green. Leaves narrow, short, erect, rolled, and brittle. Plants turn dark green and are stunted, with few tillers. Flowering delayed until late in season.	PHOSPHORUS DEFICIENCY
3. Leaves short, drooping, dark green, later with yellowing between veins and small brown blotches that coalesce from the leaf tip down. Symptoms are most severe on lower leaves. Leaves dry to a light brown. Plants stunted but tillering only slightly reduced.	POTASSIUM DEFICIENCY
4. Leaf burn at tips and along the margins of older leaves, later turn to brown elliptical spots that dry up. Plants severely stunted and eventually die.	BORON TOXICITY
5. Tips of old leaves turn brown to white. Stunted growth and reduced tillering. Salt often accumulates at the waterline on the plant stems and on soil at the edge of the field.	SALT TOXICITY

*Rare, only in areas with free calcium carbonate in the soil.
†Also sulfur deficiency, but this is rare except on new fields just developed from dry farmland.

Iron chlorosis is usually first seen when plants are 4 to 5 weeks old. The youngest leaves are most yellow, but the whole plant is often affected. The problem may be associated with iron-deficient soil but most commonly occurs in fields with highly calcareous soils and high bicarbonate levels in water. Both conditions limit the availability of iron to the plant.

Phosphorus-deficient plants initially turn yellow but rapidly change to a grayish green color with narrow, rolled, erect, brittle leaves. Lower leaves turn brown and tillering is greatly reduced, which causes the stand to appear thin. *Photo:* John F. Williams.

The yellowed and brown-blotched floating leaves on these plants are typical of zinc deficiency. The newly emerging leaves on the plant at right are green because the plant is recovering.

Rice plants suffering from boron toxicity develop brown necrotic spots along the tips and margins of older leaves and are severely stunted. *Photo:* Duane S. Mikkelsen.

Potassium-deficient plants have brown spots on their dark green leaves. Symptoms usually begin to show up late in the season.

Dead lower leaves are typical symptoms of salt toxicity. *Photo:* Duane S. Mikkelsen.

Toxic Gases and Organic Acids

Toxic substances released by the decomposition of organic matter in the flooded field may injure rice. The source of organic matter may be recently incorporated plant material or plant material that was deposited in sloughs and low areas many years ago. Under flooded conditions, decomposition of organic matter can produce various toxic organic acids through fermentation, such as lactic, butyric, acetic, and propionic acids, which later break down into toxic gases including carbon dioxide and methane. If you stir the flooded soils, gases may bubble from the soil through the water to the surface. The organic acids can sometimes be recognized by their characteristic odors.

Symptoms of toxic gas injury usually appear early in the season in fields or parts of fields where organic matter did not decompose properly before the field was flooded. The toxic acids produced accumulate in the root zone, depressing root growth and reducing water and nutrient uptake by the young plants. The roots of affected plants tend to be short, round, and stubby with reduced root hairs. They may turn brown as they die, but they do not turn black as do roots affected by hydrogen sulfide toxicity. Leaf growth is depressed, and leaves may turn dark green, then yellow; death of the leaf begins at the leaf tip and margin. Plants may die in whole areas of the field where residues remain, leaving open patches of water, or damage may be limited to retarding plant growth. Usually the greatest damage occurs in the first 30 days after flooding, but in some cases, toxicities persist longer.

If these conditions develop in your fields, drain them immediately. Exposure of the soil to the air stops the fermentation process at least temporarily and may save the stand. Allow the field to aerate for at least 7 to 14 days or until the soil starts to crack. After reflooding, carry out a plant tissue analysis for nitrogen to see if additional fertilizer is needed. Avoid incorporating crop residues in areas with a past history of toxicities.

Symptoms of toxic acid or gas injury often appear in straight, uniformly narrow areas of the field where residues were inadequately incorporated. Affected plants are often yellow, stunted, and they may die. Damage usually occurs within the first 6 weeks after stand establishment.

Hydrogen Sulfide Injury

Hydrogen sulfide injury occurs most frequently on lighter-texture soils that are relatively high in organic matter and soluble sulfates but also has been observed on heavy, finer-texture soils. Sandy to clay loam soils often have a lower cation exchange capacity and lower amounts of active iron to inactivate and precipitate sulfide ions than do clay soils. When fields are flooded under these conditions, hydrogen sulfide gradually accumulates and causes toxic reactions in rice plants. Soil from fields with high hydrogen sulfide accumulations smells like rotten eggs.

Symptoms begin to be apparent toward the end of tillering and become more severe as the plant matures. The first symptom is the change in color of the roots from white or light brown to black. In the early stages of the toxicity, the black is a film of iron sulfide that can be rubbed off. Later the whole root becomes black and begins to deteriorate. Grayish blotches appear in irregular patterns along the margins of the leaves, and the leaves begin to roll and gradually turn brown. Plants in advanced stages of hydrogen sulfide toxicity wilt, dry, and appear drought stricken, and panicles produce sterile flowers. Symptoms often occur in patchy areas of the field, reflecting differences in soil types.

Management practices for hydrogen sulfide toxicity are the same as for toxic gases and organic acids. Drain the field and allow it to aerate for at least 7 to 14 days or until the soil starts to crack. Avoid incorporating crop residues or using fertilizers containing sulfur in soils known to have a history of hydrogen sulfide toxicity problems.

If the problem is identified early, application of a nitrate or other oxidizing fertilizer onto the flooded field can partially alleviate the toxic effect of hydrogen sulfide accumulation. The addition of oxidizing materials affects oxidation-reduction changes in the soil and arrests the formation of sulfide.

The roots of plants suffering from hydrogen sulfide injury turn black; later the whole plant turns brown and dies. *Photo:* Steven C. Scardaci.

Herbicide Injury

Herbicides can injure rice when they are improperly applied to the crop, when they remain as toxic residues in the soil from a previous crop, when they are converted into a more toxic compound by soil microbes, or when they drift into fields from applications to adjacent crops. Using the wrong herbicide, applying herbicides at excessive rates, or applying them at sensitive stages in rice growth can injure rice plants seriously enough to stunt their growth or kill them. Symptoms vary according to the kind of herbicide applied. Hot temperatures and dry winds during or just after application are often associated with herbicide injury to rice, especially with applications of propanil.

Plants injured by the thiocarbamate herbicide thiobencarb initially turn blue-green and become distorted and twisted, with a slight swelling at the base of the plant. New leaves are spatulate (spoon shaped) and do not elongate. Older leaves are longer than new leaves, but older leaves are often dead or dying. Roots remain healthy.

In some soils, primarily the alluvial sandy or sandy loam soils on the east side of the Sacramento Valley, anaerobic soil bacteria break down thiobencarb into a compound that is toxic to rice plants. Affected plants are stunted and discolored. This phenomenon is called delayed phytotoxicity syndrome, or DPS.

Rice plants suffering from propanil injury have yellowing or dying leaves. Yellowing begins at the leaf tip and moves to the base of the plant. If a carbamate insecticide is applied within 14 days before or after a propanil application, injury is more severe, and plants may die.

Plants damaged by hydrogen sulfide gas appear drought stricken. Plants are usually past the tillering stage when symptoms become apparent. This field is severely damaged. *Photo:* Steven C. Scardaci.

This rice plant is suffering from thiocarbamate herbicide injury. The first leaves are dead or dying and are much longer than the short, spatulate, and distorted leaves that develop after the injury occurs. The twisted lower stem is a typical symptom. Damaged plants are usually more blue-green than normal; they may recover if the injury is not too severe.

The open spots in this field indicate areas where trifluralin residues killed rice plants. The herbicide was applied to control weeds in beans during the previous year. The damaging residues occurred at the edge of the field where the application machinery deposited heavy rates as it turned around.

Trifluralin-injured rice plants have stunted roots that often end in an enlarged knob.

This field was treated with a carbamate insecticide soon after an application of propanil. Leaves yellow and die from the tip down to the base of the plant. *Photo:* Albert A. Grigarick.

Photo: Mike Poe

Vertebrates

Rice fields provide a good habitat for various birds and rodents, some of which feed on newly planted seed, seedlings, or ripening grain. In its early development, the California rice industry was especially troubled by migrating waterfowl and blackbirds. Since the 1940s, however, both state and federal agencies have established well-situated wildlife refuges to provide alternate food sources and loafing areas for waterfowl; as a result, damage to rice by these waterfowl has been substantially reduced. Blackbird depredations continue, however, especially in early-maturing cultivars and fields in close proximity to roosting and nesting sites.

Among the rodents, muskrats and Norway rats are the most serious pests. Muskrats burrow in rice levees, damaging drainage systems and irrigation structures that regulate water flow. Major rat outbreaks occur from time to time, but even in years without outbreaks rats can damage shallow-flooded or dry-seeded rice by feeding on the germinating seed and young seedlings. They, along with house mice, mostly cause serious losses to stored rice. Beavers located in nearby flowing streams occasionally disrupt the flow of water into basins.

The object of integrated pest management programs is to alleviate damage and not necessarily to destroy the offending species. The methods must be economical, adaptable to various types of growing conditions, and create minimal hazard to people and nontarget wildlife. Plan your program and gather necessary equipment and materials in advance because immediate action is often necessary to save the crop. In many cases, you can take preventive measures to avoid damaging infestations.

The methods and chemical control materials available for vertebrate control are constantly changing. Thus, some of the methods mentioned in this manual may be outdated in a few years. Keep abreast of changes in laws and regulations concerning the status of wildlife species and the methods and materials available to control them. Your county agricultural commissioner's office is a good source for the latest information.

Waterfowl

In addition to resident waterfowl, the central valleys of California are the fall and winter home of large numbers of migratory waterfowl. These birds migrate along the Pacific Flyway to and from their northern breeding grounds and may spend from a few weeks to months annually in the state. Before the development of agriculture in the Central Valley, the vast expanses of natural marshland served as a wintering ground for millions of waterfowl as well as a brief stopover point for additional birds en route to Mexico. As California's native marshlands shrank due to agricultural and, to a much lesser extent, industrial and urban development, it was inevitable that effects on waterfowl would have to be addressed. Presently the flooding of harvested rice fields in the fall and winter provides excellent habitat with an ample food resource for ducks, geese, and other water-loving birds. During the period of flooding, rice land is often leased to hunting clubs, providing the grower with some additional income.

Among the waterfowl, ducks and geese cause the most serious losses in rice by feeding on maturing grain and sometimes causing lodging. Early-planted rice fields may attract large flocks of greater white-fronted geese, also called specks or specklebellies, which trample and bury rice seed, resulting in a thin stand. Migratory coots (mud hens and white bills), which usually prefer open water, tend to leave California in late March and do not return until after harvest in the late fall. However, some populations occasionally stay later than March and cause damage to newly planted rice fields.

An AV alarm noisemaking system frightens birds with artificially produced distresslike sounds. *Photo:* W. Paul Gorenzel.

Management Guidelines

All species of waterfowl are classed as migratory game birds protected by federal and state laws. When they are causing damage to rice fields, they cannot be killed unless a federal depredation permit is first obtained. Depredation permits are issued by the U.S. Fish and Wildlife Service upon recommendation from USDA-APHIS, Animal Damage Control (ADC). Such permits, however, are rarely issued, so growers should not count on this method for control. Game birds may be driven out of unharvested fields without a permit by shooting or with the use of exploders, shell crackers or other pyrotechnics, and sound devices, or they may be herded out with aircraft. Maturing rice can usually be protected if adequate preparations are made before the birds arrive and if frightening and herding are begun as soon as the first flocks arrive. Persistence in applying the methods and alternating frightening methods and devices are important in achieving success. Later planting may be used to avoid damage in areas where large flocks of geese are still present in the spring.

This gas cannon is not placed for maximum effectiveness. It would be more effective if elevated on a 50-gallon drum so noise would be projected across the field.

Federal and state waterfowl refuges, which provide food and water for wintering waterfowl, divert a large percentage of these potential pests away from rice fields and have been effective in reducing crop losses. Although hunting under permit is allowed on these refuges, it is delayed until most of the rice has been harvested; thus the possibility of driving birds from refuge areas to unharvested rice fields is eliminated.

The red-winged blackbird is the most commonly seen blackbird in rice fields.

Blackbirds

Blackbirds may damage ripening rice, especially during the milk and dough stages. Losses of seed and seedlings, which were common when rice fields were not flooded before planting, are not a problem in water-seeded rice in well-leveled fields, unless water is drawn off during the early developmental stages.

Five different blackbird species may damage rice. Arranged in their probable order of importance, these are red-winged blackbirds (*Agelaius phoeniceus*), tricolored blackbirds (*Agelaius tricolor*), brown-headed cowbirds (*Molothrus ater*), Brewer's blackbirds (*Euphagus cyanocephalus*), and yellow-headed blackbirds (*Xanthocephalus xanthocephalus*). Losses in certain fields, primarily those in the vicinity of large roosting areas in the Sacramento Valley, may be quite high, but only nominal losses may occur in others.

Blackbirds damage individual kernels of ripening rice.

Management Guidelines

As is the case for waterfowl management, the primary strategy for managing blackbirds is to drive them out of the field with frightening devices. Unfortunately, frightening devices are not as effective against blackbirds as they are against waterfowl. Available techniques include recorded blackbird distress calls, rifle or shotgun patrol (where it is safe), shell crackers, gas exploders, aerial pyrotechnics, and aircraft patrol. To effectively drive blackbirds away, begin these controls as soon as birds appear in the field. Combining or alternating two or more methods of control tends to increase their efficiency. Proper timing and persistence are critical. Except for tricolored blackbirds, a permit is not needed to herd or take blackbirds that are damaging or threatening to damage rice crops. The tricolored blackbird is classified by the state of California as a species of special concern. A depredation permit must be obtained before any lethal control methods can be used for tricolored blackbirds. Lethal control includes the destruction of active nests, eggs, young, or adult birds. A permit is not required to haze or scare tricolored blackbirds. Be certain of the identity of the blackbird species in the rice field before undertaking control activities. No lethal chemicals are registered for blackbird control in rice. Bird repellents for crop application are either ineffective or too costly for practical use in rice.

Bird Pests of Stored Rice

Birds are a concern around storage warehouses because their droppings and nest debris may contaminate rice. Pigeons, *Columba livia*, and English or house sparrows, *Passer domesticus*, are the birds most often involved in California. Blackbirds, *Agelaius* spp., and starlings, *Sturnus vulgaris*, are of minor importance in storage facilities, although blackbirds can be a problem in loading areas, where they feed on spilled rice. All these species can be legally controlled in and around buildings if they cause health and economic problems.

Control methods include nest destruction, shooting, live trapping, sticky roost repellents, a chemosterilant for pigeons, and the harassment chemical 4-amino pyridine. Toxic baits are generally considered only as a last resort to reduce damage. All biologically active chemical avicides are restricted materials or require a permit. For current regulations concerning bird control, consult your local agricultural commissioner; the Division of Plant Industry, California Department of Food and Agriculture; or the USDA-APHIS ADC district supervisor in your area.

Norway Rat

Rattus norvegicus

Rats may damage both stored and growing rice, and their burrowing may damage irrigation systems. They damage growing rice most seriously shortly after planting in dry-seeded fields or when the water is temporarily lowered to speed seed germination and stand establishment. Rat damage can be especially severe where fields are not level and high spots are exposed to the air. The rats pull up the sprouting plants and eat the seed. Rats may also consume ripening grain later as the cereal heads come into the milk stage, but losses at this time are generally not as serious. Burrowing by rats may cause leaks in levees and

The Norway rat is one of the most important rodent pests in rice.

Rats may nibble seed off young seedlings, killing the plants (left).

Look for rat burrows in weedy areas around the field.

around weir boxes and head gates, disrupting the irrigation system. Stored rice may be consumed by rats or contaminated by rat urine, hairs, and feces. Stringent preventive measures should be taken to exclude rats from driers and storage facilities.

The Norway rat is responsible for most rat damage in rice fields; the roof rat is rarely involved and is not generally found in association with rice culture. Norway rats are larger than roof rats and have a tail that is shorter than their head and body combined. In contrast, the roof rat's tail is longer than its head and body. Rats are mainly active at night, but if their numbers are high, some may be active during the day. They are omnivorous and feed on a wide variety of plant and animal materials; however, the roof rat shows a preference for soft fruits and nuts. Norway rats are good diggers and burrow along foundations of buildings, beneath rubbish piles, in the banks along canals, and in levees and checks of rice fields.

Both rat species are prolific breeders; an average of about 20 young are weaned per year by a female in the wild. Both rats are capable of breeding year-round; however, in California, there are two distinct breeding periods: spring and fall. Little reproduction generally occurs during the heat of the summer and the cool of the winter. Rat mortality is high, and most do not live longer than a year.

Management Guidelines

Habitat Management. Both rats and muskrats favor marshlike habitats where water is available year-round. They usually invade rice fields from their natural habitat of rivers, creeks, and canals. Infestations in rice can be prevented or slowed by keeping rat and muskrat populations at low levels in natural habitats adjacent to rice fields.

Keeping weed growth mowed or grazed along levees and controlling weeds along the banks of canals and ditches make them less suitable as rat habitat and help keep rat populations down. Rotation with tomatoes or cereal crops, which are not good rat habitat, will drastically reduce rat populations. Periodic crop rotation or fallowing will significantly reduce the magnitude, frequency, and speed of rat reinvasion once the field is returned to rice.

Laser leveling and the shift to larger rice checks has contributed much toward curtailing rat damage by reducing the length and number of levees and increasing the size of basins. The decrease in levees significantly limits places for rats to burrow and live.

Baiting. As nongame mammals, Norway and roof rats may be taken at any time or in any manner when they are injuring growing crops or other property. Various forms of baits are used to control them. Consult your agricultural commissioner or farm advisor for specific information.

Broadcast or spot baiting with a grain-based bait formulated with an acute rodenticide such as zinc phosphide (a restricted-use pesticide) is generally considered the most economical of the rodenticides for field rat control. Broadcast baiting, along with rate of application, is specified on some bait product labels. Such application, whether carried out by hand, with a mechanical spreader, or by aircraft, may not be as effective as a thorough spot-baiting treatment. However, it can be carried out more rapidly and is usually the more economical method of application, especially where infestations are large. Rat infestations are best controlled prior to planting, as no rodenticide is registered for application directly onto the crop. Spot baiting is accomplished by placing a small quantity of bait according to label instructions in or just in front of each active burrow.

Control is significantly improved by prebaiting with clean (nontoxic) grain several days prior to applying the toxic bait. This conditions the rats to feed on this new food resource so that they will more readily accept and consume the toxic bait when applied. Such prebaiting increases the likelihood that rats will be eager eaters of the acute toxic bait, resulting in a higher level of mortality. Rats surviving the initial toxic bait treatment will likely be bait shy as the result of consuming a sublethal dose of the bait. These bait-shy rats will refuse to eat the same zinc phosphide bait if it is offered again too soon after the initial treatment. Therefore, it is recommended to wait at least 3 months, and preferably 6 months, before treating again with zinc phosphide bait. The bait shyness problem can be circumvented by switching to a slow-acting anticoagulant bait. This allows an immediate follow-up application of anticoagulant bait to control any surviving rats.

Two anticoagulant rodenticides, diphacinone and chlorophacinone, are registered for field rodent control. Although more expensive, these slow-acting rodenticides, which require multiple feeding to be effective, give excellent control and do not produce bait shyness. Methods of application may include spot or broadcast baiting or the placement of bait in enclosed stations (bait boxes). Be sure the bait label permits your intended use and method of application. Bait stations are a common and effective method of applying anticoagulant bait because such stations provide the needed repeat feedings and can be serviced and maintained over prolonged periods. Depending on the size of the rat infestation, a series of bait stations will be needed and are generally placed 50 to 100 feet (15.2–30.5 m) apart. A satisfactory station can be made from 3- or 4-inch (7.6–10.2 cm) PVC pipe or sections of automobile tires, although commercially made stations may be purchased. Stations shelter bait from inclement weather, exclude nontarget species, and afford seclusion to the feeding rats.

Protection of Stored Rice. Rodent-proofing the building is the best and only long-term solution. Rodent proofing should be aimed at house mice as well as rats. The most common preventive measure for a potential rodent problem in stored rice is to maintain a series of permanent outdoor bait stations containing anticoagulant bait. Place the stations 50 to 100 feet (15.2–30.5 m) apart at strategic places around the exterior of the building to control rats and house mice before they enter the storage facility. This is commonly referred to as perimeter baiting. Special care must be taken inside the storage installation not to contaminate the rice with rodenticides. For rodent control in and around storage facilities, anticoagulant baits are generally preferred over other rodenticides. Where rodenticide contamination of food is a risk or when only a few animals are present, use traps. Wooden snap traps are inexpensive and very effective.

You can make a bait station of corrugated plastic tubing or other materials that will shelter bait and prevent feeding by game birds and other wildlife.

Perimeter baiting for rats can be supplemented with burrow fumigation if needed. Several different kinds of cartridges are available for fumigating rodent burrows. When ignited, these cartridges produce suffocating gas. Fumigation cartridges are available from many county agricultural commissioners' offices and from retail outlets. No permit is required for their use, but be sure to follow all label directions. Fumigation is most effective when the soil is moist. Use fumigation cartridges when freshly excavated soil indicates that the burrow is active. After shoving the ignited cartridge into a burrow, seal the burrow opening with soil. Wait 24 hours and re-treat burrows that have been reopened. Do not use smoke cartridges where a fire hazard exists or for burrows that are near or underneath buildings. Fumigation works well for removing rats that have burrowed under or along a rail siding.

Muskrat

Ondatra zibethica

Muskrats are semiaquatic rodents that sometimes inhabit natural waterways, water supply canals, and drainage ditches near rice fields. Their burrowing, especially around head gates, can cause breaks in levees and dikes and subsequent loss of water from rice basins (Fig. 20). Repairs are costly, and significant yield may be lost before repairs can be completed. Muskrats also occasionally cut and eat rice plants.

Muskrats are ratlike in shape; have a rich chocolate brown fur; a long, laterally compressed tail; and partly webbed feet. Adults are about 18 inches (45 cm) long and weigh from 1½ to 2½ pounds (0.7–1.1 kg). Muskrats are native to North America and commonly feed on cattails, tules, and the roots of bermudagrass and other plants. They also eat crayfish, and their burrow entrances are often strewn with crayfish remains. The peak breeding season for muskrats is in the spring, but young may be born throughout the year. Fewer births occur in winter than in any other season. Muskrats are most active at night.

Management Guidelines

A number of management methods are available to reduce losses from muskrats. Mowing or using herbicides to control weed growth on levees can help make the habitat less suitable for both muskrats and rats, and make it easier to detect burrows for monitoring and control purposes. You can prevent damage to canal head gates and levee boxes by constructing them of concrete or metal-lined wood with galvanized metal wings.

Trapping is effective for removing muskrats because generally their numbers are low. Experienced trappers can usually trap muskrats with ease. Conibear traps (size 110) and cage-type live traps are both suitable for taking muskrats. Place Conibear traps over burrow entrances and secure them to a stake; they are triggered when the muskrat emerges from the burrow. Live traps are most effective when baited and placed near burrow entrances or along well-used pathways.

Anticoagulant baits are effective and are commonly placed in bait stations accessible to the muskrats. Bait stations can be located adjacent to burrow entrances. Paraffin-type anticoagulant baits are especially useful because they are resistant to mold induced by moisture and high humidity. Floating bait stations or boxes with large Styrofoam bases are also very effective where the habitat will accommodate them. Muskrats like to climb onto floating objects and will readily enter the bait station and feed. To achieve control, keep the stations well supplied with bait until all feeding stops.

At present, acute rodenticides such as zinc phosphide are not recommended to control muskrats in California. Burrow fumigants are very effective for muskrat control, but very few are currently registered in California. Fumigants are easier to apply when burrows are exposed; so where it is possible, especially in drainage ditches, lower the water level to expose the burrow entrances. After applying the fumigant, seal the opening off with mud to hold a lethal concentration of gas in the burrow system.

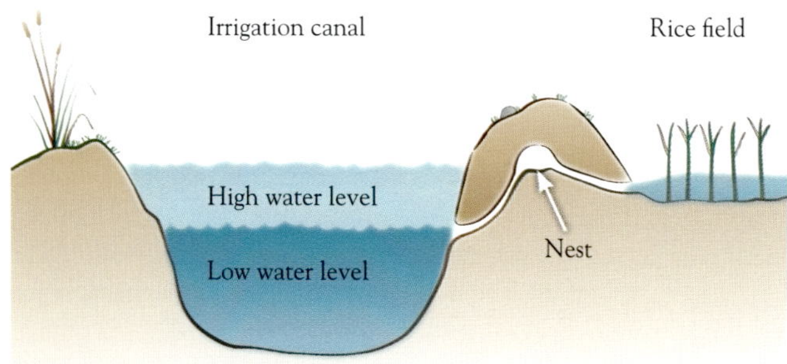

Figure 20. Cross section of an irrigation system damaged by muskrat burrowing.

Adult muskrat. *Photo:* Desley Whisson.

Suggested Reading

General

California Rice Production Workshop Workbook. Updated annually. Published by University of California Cooperative Extension and the California Rice Research Board.

Cultivated Wild Rice Production in California. 2007. D. Marcum. UC ANR Publication 21622.*

Feeding Rice Straw to Cattle. 2002. D. Drake, G. Nader, and L. Forero. UC ANR Publication 8079.*

Rice Production in California. 1992. J. E. Hill. UC ANR Publication 21498.*

Rice Quality Handbook. 2009. R. G. Mutters and J. F. Thompson. UC ANR Publication 3514.*

Rice Straw for Use in Dairy Heifer Rations. 2010. G. A. Nadar and P. H. Robinson. UC ANR Publication 8392.*

Integrated Pest Management

IPM in Practice: Principles and Methods of Integrated Pest Management. 2nd ed. 2012. M. L. Flint. University of California Statewide Integrated Pest Management Program. UC ANR Publication 3418.*

UC IPM Pest Management Guidelines: Rice. Revised continuously. University of California Statewide Integrated Pest Management Program. UC ANR Publication 3465.* Also available from University of California Cooperative Extension offices and at the UC IPM website, http://www.ipm.ucdavis.edu.

Soil, Water, Weather, and Nutrients

Agricultural Salinity and Drainage. 1999. B. Hanson. UC ANR Publication 3375.*

CIMIS: California Irrigation Management Information System. Online at http://www.cimis.water.ca.gov/.

Diagnosing Soil Physical Problems. 1976. W. E. Wildman. UC ANR Publication 2664.*

Irrigation Water Salinity and Crop Production: Farm Water Quality Planning Series. 2002. S. Grattan. UC ANR Publication 8066.*

Rice Irrigation Systems for Tailwater Management. 2nd ed. 1995. J. E. Hill. UC ANR Publication 21490.*

Seepage Water Management: Voluntary Guidelines for Good Stewardship in Rice Production. 1998. UC ANR Publication 21568.*

Soil and Plant Tissue Testing in California. 1983. H. M. Reisenauer, ed. UC ANR Publication 1879.*

Western Fertilizer Handbook. 9th ed. 2002. Western Plant Health Association. Interstate Publishers, Danville, IL.

Pesticide Application and Safety

La loteria de los pesticidas. 1992. M. L. Flint. University of California Statewide Integrated Pest Management Program. UC ANR Publication 3355.*

National Pesticide Safety Education Core Manual. 2005. C. Randall et al., eds. U.S. Environmental Protection Agency, Office of Pesticide Programs, Washington, D.C.

Pesticide Safety: A Reference Manual for Private Applicators. 2nd ed. 2006. P. O'Connor and S. Cohen. University of California Statewide Integrated Pest Management Program. UC ANR Publication 3383.*

Pesticide Safety for Small Farms. 1996. UC ANR Publication 6565D (DVD: English and Spanish), UC ANR Publication 6566D (DVD: Hmong, Lao, Ilokano). Also available in VHS formats.*

Pesticide Safety Information Series. Rev. 2003. California Department of Food and Agriculture, Sacramento, CA. Includes specific procedures for handling hazardous pesticides.

Residential, Industrial, and Institutional Pest Control. 2nd ed. 2006. P. O'Connor-Marer. University of California Statewide Integrated Pest Management Program. UC ANR Publication 3334.*

Safe and Effective Use of Pesticides. 2nd ed. 2000. P. Marer. University of California Statewide Integrated Pest Management Program. UC ANR Publication 3324.*

Safe Handling of Pesticides. 1988. UC ANR Publication V88-T (VHS), UC ANR Publication 6518D (DVD: English and Spanish).*

Seguridad en el manejo de pesticidas. 2nd ed. 2007. P. O'Connor-Marer and S. Cohen. University of California Statewide Integrated Pest Management Program. UC ANR Publication 3394.*

El uso seguro de los pesticidas. 1988. UC ANR Publication 6518D (DVD).*

Insects

California Insects. 1979. J. A. Powell and C. L. Hogue. University of California Press, Berkeley, CA.

Insect Identification Handbook. 1984. C. A. Ferris. UC ANR Publication 4099.*

Insect Pests of Farm, Garden, and Orchard. 7th ed. 1979. R. H. Davidson and W. F. Lyon. Wiley, New York.

Introduction to the Study of Insects. 6th ed. 1989. D. J. Borror, C. A. Triplehorn, and N. F. Johnson. Saunders College Publishing, Philadelphia.

National Audubon Society Field Guide to North American Insects and Spiders. 1995. L. Milne and M. Milne. A. A. Knopf, New York.

Natural Enemies Handbook: The Illustrated Guide to Biological Pest Control. 1998. M. L. Flint and S. H. Dreistadt. University of California Statewide Integrated Pest Management Program. UC ANR Publication 3386.*

Stored-Grain Insects. U.S. Department of Agriculture Handbook 500. Washington, D.C.**

Diseases and Nematodes

Compendium of Rice Diseases. 1992. R. K. Webster and P. S. Gunnell, eds. American Phytopathological Society, St. Paul, MN.

Plant Pathology. 5th ed. 2005. G. N. Agrios. Elsevier Academic Press, Burlington, MA.

Rice Diseases. 1972. S. H. Ou. Commonwealth Mycological Institute, Kew, Surrey, U.K.

Weeds

Applied Weed Science. 1985. M. A. Ross and C. A. Lembi. Burgess Publishing Co., Minneapolis, MN.

Aquatic and Riparian Weeds of the West. 2003. J. M. DiTomaso and E. A. Healy. UC ANR Publication 3421.*

How to Identify Plants. 1957. H. D. Harrington and L. W. Durrell. Sage Press, Denver, CO.

Weeds of California and Other Western States. 2007. J. M. DiTomaso and E. A. Healy. UC ANR Publication 3488.*

Weed Science. 3rd ed. 1996. W. P. Anderson. West Publishing, Minneapolis-St. Paul, MN.

Vertebrates

Rats Pest Note. 2003. T. P. Salmon, R. E. Marsh, and R. M. Timm. University of California Statewide Integrated Pest Management Program. UC ANR Publication 74106.***

Vertebrate Pest Control Handbook. 4th ed. 1994. J. P. Clark, ed. California Department of Food and Agriculture, Sacramento, CA. (Also online at http://www.vpcrac.org/about/handbook.php.)

Wildlife Pest Control Around Gardens and Homes. 2nd ed. 2006. UC ANR Publication 21385.*

---

*Available from University of California, Agriculture and Natural Resources, Communication Services, 1301 S. 46th Street, Building 478 – MC 3580, Richmond, CA 94804-4600 (http://anrcatalog.ucanr.edu).

**Available from Superintendent of Documents, U.S. Government Printing Office, Washington D.C. 20402.

***Available from the UC ANR website, http://anrcatalog.ucanr.edu, and from the University of California Statewide Integrated Pest Management Program website, http://www.ipm.ucdavis.edu, under "Home, Garden, Turf & Landscape Pests."

Glossary

abiotic disorder. a disease caused by factors other than a pathogen; physiological disorder.

acid, acidic. having a pH less than 7.

adventitious. a structure arising from an unusual place, such as roots growing from leaves or stems.

alkaline. basic, having a pH greater than 7.

annual. a plant that normally completes its life cycle of germination, growth, reproduction, and death in a single year.

anticoagulant. a substance that prevents blood clotting, resulting in internal hemorrhaging; may be used as a rodenticide.

auricles. the earlike projections at the base of the leaf blades of some grasses; used to identify species (see Fig. 11).

awn. a slender, bristlelike organ usually at the apex of a plant structure.

axil. the upper angle between a branch or petiole and the stem from which it grows.

basin. a portion of the rice field bounded by levees.

biotic disease. a disease caused by a pathogen, such as a bacterium, fungus, phytoplasma, or virus.

biotype. a strain of a species that has certain biological characteristics separating it from other individuals of that species.

boot. a bulge in the upper leaf sheath caused by the expansion of the developing panicle.

borrow pits. depressions on either side of a levee created when soil is removed from the field to build the levee (see Fig. 10).

calyx. the sepals of a flower; they enclose the unopened flower bud.

canker. a dead, discolored, often sunken area (lesion) on a root, stem, or stolon.

caterpillar. the larva of a butterfly, moth, sawfly, or scorpionfly.

check. the part of a rice field between two levees; same as basin.

chlamydospore. overwintering reproductive spore of a smut fungus.

chlorophyll. the green pigment of plants that captures the energy from sunlight necessary for photosynthesis.

chlorosis. yellowing or bleaching of plant tissue that is normally green, usually caused by the loss of chlorophyll.

coleoptile. a sheathlike structure enclosing the shoot of a grass seedling (see Fig. 4).

collar region. in grasses, the region where the leaf blade and sheath meet; used in identifying species (see Fig. 11).

conidium (plural, conidia). an asexual fungus spore formed by fragmentation or budding at the tip of a specialized hypha.

control action guideline. a guideline used to determine whether pest control action is needed.

cotyledons. the first leaves of the embryo formed within a seed and present on seedlings immediately after germination; seed leaves (see Fig. 11).

cross-resistance. in pest management, resistance of a pest population to a pesticide to which it has not been exposed that accompanies the development of resistance to a pesticide to which it has been exposed.

culm. the jointed stem of a grass.

cultivar. a variety or strain developed and grown under cultivation.

disease. a disturbance of a plant that interferes with its normal structure, function, or economic value and is typically manifested by signs or symptoms.

dough stage. a stage in grain development when the grain turns from a liquid to a soft, doughy consistency before hardening.

drift. the dispersal of a substance, usually a pesticide, beyond the bounds of its intended application.

economic threshold. a level of pest population or injury at which the cost of a control action equals the loss of crop value prevented by that control action.

epidermis. the outermost layer of living cells on the surface of a plant or animal.

evapotranspiration. the loss of soil moisture by the combination of soil surface evaporation and transpiration by plants.

flag leaf. the last emerging leaf below the panicle of a grass (see Fig. 6).

floret. the individual flower of a grass spikelet.

host. a living organism that is invaded by a parasite and from which the parasite obtains its food.

hyphae. the filaments that make up the body of a fungus.

immune. not susceptible or responsive to a given pathogen; having a high degree of resistance to a disease.

infection. the entry of a pathogen into a host and establishment of the pathogen as a parasite of the host.

infestation. the presence of a large number of pest organisms in an area or field, on the surface of a host or anything that might contact a host, or in the soil.

inflorescence. flower cluster.

inoculum. any part or stage of a pathogen, such as spores or virus particles, that can infect a host.

instar. an insect between successive molts; the first instar is between hatching and the first molt.

internode. the area of a stem between nodes.

invertebrate. an animal having no internal skeleton.

jointing. elongation of rice internodes before flowering.

larva (plural, larvae). the immature form of an insect, such as a caterpillar or maggot, that hatches from an egg, feeds, then enters a pupal stage.

lesion. a localized area of diseased tissue, such as a canker or leaf spot.

ligule. in many grasses, a short membranous projection on the inner side of the leaf blade at the junction where the leaf blade and leaf sheath meet (see Fig. 11).

lodging. the toppling of plants of a grain crop before harvest, often caused by wind, rain, or waterfowl.

metamorphosis. a change in form during development.

microorganism. an organism of microscopic or small size.

milk stage. the early stage of grain development when the grain is filled with a milky fluid.

mutation. the abrupt appearance of a new, heritable characteristic as the result of a change in the genetic material of one individual cell.

mycelium (plural, mycelia). the vegetative body of a fungus, consisting of a mass of slender filaments called hyphae.

natural enemies. predators, parasites, or pathogens that are considered beneficial because they attack and kill organisms that we normally consider to be pests.

necrosis. death of tissue accompanied by dark brown discoloration, usually occurring in a well-defined part of a plant, such as the portion of a leaf between leaf veins or the xylem or phloem in a stem.

node. the slightly enlarged part of a stem where buds are formed and where leaves, stems, and flower trusses originate.

nymph. the immature stage of insects such as lygus bugs and aphids that gradually acquire adult form through a series of molts without passing through a pupal stage.

panicle. a branching cluster of flowers held on a stem, such as the inflorescences of most grasses.

parasite. an organism that lives in or on the body of another organism (the host) from which it derives its food without killing the host directly; also used in this manual to describe an insect that spends its immature stages in the body of a host, which is killed just before the parasite pupates.

pathogen. a disease-causing organism.

perennial. a plant that can live 3 or more years and flower at least twice.

petiole. the stalk connecting a leaf or leaflets to a stem.

pH. a value used to express relative acidity or alkalinity.

phloem. the food-conducting tissue of a plant vascular system.

photosynthesis. the process whereby plants use light energy to form sugars and other compounds needed to support growth and development.

phytotoxicity. the ability of a material such as a pesticide or fertilizer to cause injury to plants.

pistil. female part of the flower, usually consisting of ovules, ovary, style, and stigma.

postemergence herbicide. an herbicide applied after target weeds emerge.

predator. an animal that attacks and feeds on other animals (the prey), usually eating most or all of the prey and consuming many prey during its lifetime.

preemergence herbicide. an herbicide applied before target weeds emerge.

primary inoculum. the initial source of a pathogen that starts disease development in a given location.

propagule. any part of a plant from which a new plant can grow, including seeds, bulbs, rootstocks, etc. Also, any structure of a pathogen that can serve as a source of inoculum.

protectant fungicide. a fungicide that protects a plant from infection by a pathogen.

pupa (plural, pupae). the nonfeeding, inactive stage of an insect between larva and adult in insects with complete metamorphosis.

reservoir. the site where a pest population or quantity of inoculum can survive in the absence of a host crop, and from which a new crop may be invaded.

residue management. management of rice straw and stubble after harvest.

resistant. able to withstand conditions harmful to other strains of the same species.

respiration. the process by which nutrients are metabolized to provide energy needed for cellular activity.

rhizome. a horizontal underground stem, especially one that roots at the nodes to produce new plants (see Fig. 11).

rootstock. an underground stem or rhizome.

sanitation. any activity that reduces the spread of pathogen inoculum, such as removal and destruction of infected plant parts and cleaning of tools and field equipment.

sclerotium (plural, sclerotia). a compact mass of hardened mycelium that serves as a dormant stage for some fungi.

secondary pest outbreak. the sudden increase of a pest population that is normally at low or nondamaging levels caused by the destruction of natural enemies by treatment with a nonselective pesticide to control a primary pest.

secondary spread. the spread of a pathogen within a field after the initial or primary infection.

seed leaf. leaf formed within a seed and present on a seedling at germination; cotyledon.

selective pesticide. pesticide that is more toxic to target pests than to natural enemies.

senescent. growing old; aging.

sheath. the part of a grass leaf that encloses the stem below the collar region (see Fig. 11).

spikelets. the small groups of flowers that make up the panicles of grasses.

spiracle. an external opening of the system of ducts, or tracheae, that serves as a respiratory system in insects.

sporangium (plural, sporangia). a structure containing asexual spores.

spore. a reproductive body produced by certain fungi and other organisms, capable of growing into a new individual under proper conditions.

stolon. a stem that grows horizontally along the surface of the ground.

sun checking. cracking or breaking of whole kernels of grain caused by exposure to alternating conditions of dew, sun, and water stress.

systemic. capable of moving throughout a plant or other organism, usually in the vascular system.

terrestrial biotype. a strain of an organism adapted to growing on land rather than in water.

tiller. a young vegetative shoot in grasses (see Fig. 5).

tolerant. a cultivar that is able to grow and produce an acceptable yield when infected by a pathogen.

toxin. a poisonous substance produced by a living organism.

translocated herbicide. herbicide that is able to move throughout a plant after being applied to leaf surfaces.

transpiration. the evaporation of water from plant tissue, mostly through stomata.

treatment threshold. the level of a pest population, usually measured by a specified monitoring method, at which a control measure is needed to prevent eventual economic injury to the crop.

true leaf. any leaf produced after the seed leaves (cotyledons).

tuber. an enlarged, fleshy, underground stem with buds capable of producing new plants.

variety. an identifiable strain within a species, usually referring to a strain which arises in nature as opposed to a cultivar, which is specifically bred for particular properties; sometimes used synonymously with cultivar.

vascular system. the system of plant tissues that carries water, mineral nutrients, and products of photosynthesis throughout the plant, consisting of xylem and phloem.

vector. an organism capable of transmitting a pathogen to a host.

vegetative growth. growth of leaves, roots, and stems, including tubers, as opposed to flowers or fruits.

virulent. capable of causing a severe disease; strongly pathogenic.

xylem. tissue that conducts water and mineral nutrients from the roots to the rest of the plant.

Y-leaf. the most recently matured leaf.

zoospore. a motile spore.

Index

Illustrations and photographs are indicated with *f*. Tables are indicated with *t*.